APPLIED PHYSIOLOGY OF EXERCISE

APPLIED PHYSIOLOGY OF EXERCISE

Dr Govindasamy Balasekaran
Nanyang Technological University, Singapore

Dr Visvasuresh Victor Govindaswamy
Concordia University, USA

Lim Ziyuan Jolene
Nanyang Technological University, Singapore

Boey Peck Kay Peggy
Nanyang Technological University, Singapore

Ng Yew Cheo
Singapore University of Social Sciences, Singapore

World Scientific

NEW JERSEY • LONDON • SINGAPORE • BEIJING • SHANGHAI • HONG KONG • TAIPEI • CHENNAI • TOKYO

Published by

World Scientific Publishing Co. Pte. Ltd.
5 Toh Tuck Link, Singapore 596224
USA office: 27 Warren Street, Suite 401-402, Hackensack, NJ 07601
UK office: 57 Shelton Street, Covent Garden, London WC2H 9HE

Library of Congress Cataloging-in-Publication Data
Names: Balasekaran, Govindasamy, author.
Title: Applied physiology of exercise / Dr. Govindasamy, Nanyang Technological University, Singapore Balasekaran [and four others].
Description: Hackensack : World Scientific, 2022. | Includes bibliographical references and index.
Identifiers: LCCN 2021013059 | ISBN 9789811232787 (hardcover) | ISBN 9789811234156 (paperback) | ISBN 9789811233753 (ebook) | ISBN 9789811233760 (ebook other)
Subjects: LCSH: Exercise--Physiological aspects.
Classification: LCC QP301 .B312 2022 | DDC 612.7/6--dc23
LC record available at https://lccn.loc.gov/2021013059

British Library Cataloguing-in-Publication Data
A catalogue record for this book is available from the British Library.

For any available supplementary material, please visit
https://www.worldscientific.com/worldscibooks/10.1142/12174#t=suppl

Desk Editor: Lai Ann

Typeset by Stallion Press
Email: enquiries@stallionpress.com

Dedication

This book is specially dedicated to all Physical Education teachers, coaches, and sports and health practitioners. The authors would like to thank everyone who has helped in one way or another to make this publication possible. We will run out of space if we thank everyone. Nonetheless, each one of you will always be in our hearts as you have selflessly helped as a true colleague and friend. Special thanks to all the students who were taught physiology by us and enjoyed it as much as we did. Lastly, a special mention to Ms Yow Chea Nuan for her tremendous support and help to make this publication a reality.

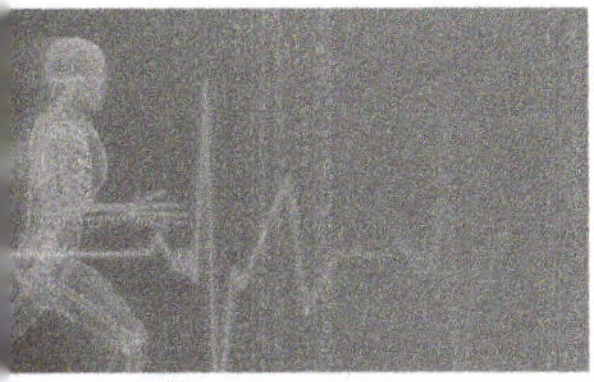

Foreword

Stephen L. Kopecky, MD
Professor of Medicine
Department of Cardiovascular Diseases
Mayo Clinic
Rochester Minnesota, United States of America (USA)

Director, Preventive Cardiology Fellowship
Past President, American Society of Preventive Cardiology
Author, *Live Younger Longer* (Available September 2021)

I am a preventive cardiologist at Mayo Clinic in Rochester Minnesota USA and have both ordered and interpreted thousands of oxygen consumption exercise tests on my patients over the past four decades. Exercise physiology is something I use every day in my work as I try to put it into terms that patients can understand in order to improve their fitness, health, and longevity. Exercise physiology is a dynamic field with many new discoveries such as high-intensity interval training, which has helped us learn how to improve our fitness and performance and physically push our bodies to new heights.

It has been a great pleasure for me to work over the past decade with The Foundation for Global Community Health efforts to increase the understanding and enjoyment of regular physical activity for children worldwide. In this context I have been able to interact and learn from the lead author of *Applied Physiology of Exercise*, Dr Govindasamy Balasekaran. Dr Bala, as he is called, is a true expert not just in the physiology of exercise but also in his ability to convey key concepts of complex human physiology in terms that are understandable for all. I remember discussing and learning from Dr Bala the interval training techniques of the "Flying Finn", Paavo Nurmi, which enabled him to win multiple Olympic gold medals in distance running. This had particular interest to me since Dr. Roger Bannister, a 4th year medical student without much time to train, utilized similar interval training techniques before he became the first human to run the mile in less than 4 minutes.

Dr Bala's experience in learning, researching, and teaching human physiology shows through very well in this book. His ability to not only know and understand the key concepts but also to apply that knowledge is present in each chapter. The book includes key questions at the end of each chapter which have likely been asked of, and by, Dr Bala hundreds of times. These questions help the reader focus on the essential knowledge needed to understand and apply these complex physiologic mechanisms to an individual athlete.

Applied Physiology of Exercise is well organized and helps the reader learn the basics of human physiology in the early chapters including ATP systems, anaerobic glycolysis, and aerobic metabolism and then incorporate these key concepts in chapters on oxygen uptake and adaptations. Numerous illustrations synopsize salient points while multiple appendices allow the reader to utilize these mechanisms to assess and optimize an athlete's training and physical performance.

In conclusion, an understanding of human muscle physiology, energy production and utilization, including oxygen uptake, and ways to improve it are essential for all who coach, train, educate, or deliver care to athletes at any level. With the world becoming a very busy place, we must help not only our athletes but also our students, our patients, and our colleagues understand the most efficient way to help our bodies maintain cardiovascular conditioning. This book discusses these key concepts and helps the reader learn how to achieve these goals. For me, Dr Bala and this text have helped my understanding of human physiology and how to apply it for my patient's benefit. Whatever area you are in, be it educating, coaching, training, or caregiving, it is likely that both you and those you work with will benefit from this book also.

Theodore J. Angelopoulos, Ph.D., MPH., FACSM, FTOS

Professor and Chair
Department of Rehabilitation and Movement Sciences
University of Vermont
Burlington, Vermont, United States of America (USA)

Dr G. Balasekaran and his co-authors have a unique perspective on presenting key concepts of applied physiology of exercise coupled with applications in this book.

The book provides a thorough yet succinct introduction to the basic principles of applied physiology as it pertains to exercise and sports. Its applied perspective is designed to help future physical educators and exercise professionals understand and appreciate the scientific foundations of exercise and sport. Written in a clear and easy to understand style, this book will be a useful resource for a variety of applications, including developing plans for physical education classes, organizing effective practice sessions for athletes and prescribing safe and effective exercise programs for health promotion.

The authors provide an essential introduction to the energy systems before covering important aspects of exercise and sport physiology. The text is accompanied with high quality figures and very useful tables.

This book will be a useful resource for students as they learn to become physical educators, coaches and exercise professionals. It provides the right amount of practical information they will need to apply in schools, athletic organizations, health clubs and hospitals.

In summary, the applied physiology of exercise book is a very valuable resource. It really helps the students to understand the biological demands to perform various types of activities, the adaptive responses to physiological stressors and the inherent biological ability to improve physiological capacity.

Preface

Govindasamy Balasekaran, Ph.D., FACSM

Associate Professor
Physical Education & Sports Science
National Institute of Education
Nanyang Technological University
Singapore

Health Fitness Director™
American College of Sports Medicine (ACSM)

President
Asian Council of Exercise & Sports Science (ACESS)

There are many coaches, personal trainers, teachers, and health practitioners who are training individuals to improve fitness or elite performances. Previously, trainings were formulated by the trial and error method, which has a high probability of hit and misses, therefore non-advisable, whereas current training methods are credible as it is evidence based through science. That being said, training has revolutionized.

A long time ago, in the 1920s, marathoners used to eat a steak before they run a marathon as they want to have the strength and endurance of the bull. Paavo Nurmi, the 'Flying Finn' or the 'Phantom Finn', dominated distance running in the early 20th century. With his intense workouts, Nurmi set 22 official world records and won nine gold medals in the Olympic games at distances between 1500 metres and 20 kilometres; Arthur Lydiard, a New Zealand coach, the master of long slow distance running, made his runners run heavy mileage for many weeks; the great Czechoslovakian, Emil Zatopek, whose unorthodox training at high volume with high-intensity intervals and voluminous mileage had won himself three gold medals in the 5,000 m, 10,000 m, and marathon in one Olympic Game; and Lasse Artturi Viren, who trained thrice a day, earned himself four gold medals in the 5,000 m and 10,000 m races during two successive Olympic Games (1972 and 1976). In contrast, a recently established athlete, Mohamed Muktar Jama Farah (Mo Farah) accomplished the same feat as Lasse Viren, but was science-based knowledge incorporated into his high-quality training programme? This book will explore and give some answers to his training regime. Moreover, the world's fastest men in sprints and marathon, Usain Bolt and Eliud Kipchoge, respectively, had intelligent training taxing the various components of fitness in combination with added strength workouts, which led them to being current record holders in their respective distances.

In the current age, the world also recognises middle-distance specialists, Sebastian Newbold Coe and Wilson Kosgei Kipketer, as legends in their own times but with different training philosophies to attain the same goal of superior performances. Both of them had world records which stood the test of time. Footballers like Cristiano Ronaldo and Lionel Messi, and rugby player like Tendai Mtawarira, nicknamed 'the beast', were fit beyond any ordinary individual athlete. How did they achieve such fitness status and how much training is considered too much? It is understood that 'more' is unnecessary as 'diminishing returns' are one of the many physiological adaptations. In addition, consistent large volume and high-intensity training increases the susceptibility to permanent injuries.

Training has to be science-based and science is the only way forward, thus the book title indicates *Applied Physiology of Exercise*. This book is suitable for anyone interested in training with the use of science. Any training can be answered with physiological rationale. If it cannot be answered, we are moving away from specific intelligent training and into erroneous combination high load training (for example, combining aerobic and anaerobic interval training in a single training session) that may not elicit a higher percentage of physiological adaptations but may induce injuries as the body is not conditioned properly. Combination high-level training may be introduced at a later stage once an individual goes through specific conditioning following a general conditioning of building the 'base' or 'foundation' period for at least six months. Factual training with science takes time to attain superior performance without performance-enhancing drugs or supplements such as growth hormones and testosterone.

In addition to training, what about weight loss programmes for children and adults? High-intensity interval training has been an increasingly popular training method for weight loss. However, would this method be sustainable in a long run? This book also teaches how to incorporate the utilisation of fat during exercise sessions. Science-based knowledge is essential to have for the right successful long-term weight loss programmes.

Hence, through this book, we are advocating training with science-based facts and eliminating the use of performance-enhancing drugs or supplements. The latter could be the root cause of diseases and sickness in later life. Shortcut to training is only short-term glory and quality of life is the essence. Gathering the 'right' knowledge is important and hopefully readers will be better equipped after reading this book. There are questions in each chapter to enhance learning and comprehension. It requires readers to think, rationalize, answer, and apply the facts to training or weight loss programmes. These questions aim to ignite the critical component of learning as readers critique and re-analyse their training programme. Even though each training could be different with everyone holding on to a different training philosophy, facts through science are universal for all.

I had warmly invited Professor Kopecky and Professor Theodore (Ted), both who are dear friends to me whom I have known for a long time, to write their foreword for this textbook. Prof Kopecky is a world-renowned expert especially in the area of cardiology. He was the past President of American Society of Preventive Cardiology and a Professor of Medicine in the Mayo Clinic, USA. Prof Ted is a Professor and Chair of the Department of Rehabilitation and Movement Sciences, University of Vermont, USA and is also a Fellow of the American College of Sports Medicine (FACSM). With their valuable insights, I am honored to have both of them write the foreword for us.

I hope all of you will have a wonderful time reading and reap the benefits of gaining new knowledge in training for yourself or others. Training can be fun once it is re-focused on quality instead of quantity. Intelligent science-based training will make an individual healthy and injury free as undisrupted consistent training will make one fitter and faster. Be patient as you train with science as it will take a long time to help you or your student(s)/athlete(s) reach towards the goal as a healthy individual or champion!

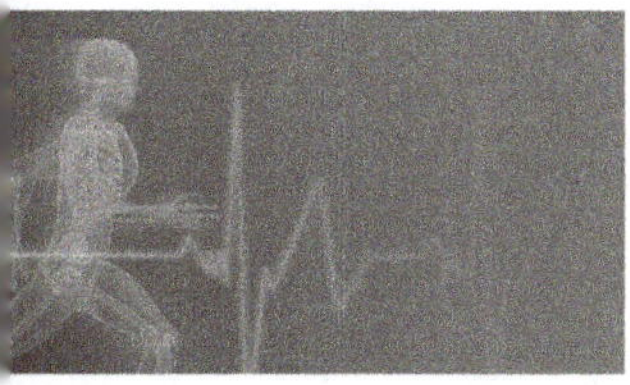

About the Authors

Main Author

Dr Govindasamy Balasekaran, better known as Dr Bala, earned his PhD at the University of Pittsburg under the tutelage of Professor Robert J. Robertson, famous worldwide for his rate of perceived exertion work in exercise, and Dr Silva Arslanian, MD, a well-known pediatric endocrinologist from the University of Pittsburgh Medicine Center, Children's Hospital of Pittsburgh, USA. Following which, he completed his Post-Doctoral Fellowship in molecular genetics at the University of Pittsburgh with Dr Robert E. Ferrell, a world-renowned genetics professor. Dr Bala is the former Head of Physical Education & Sports Science and former Programme Director of Sport Science and Management at the Physical Education and Sports Science department in the National Institute of Education (NIE), Nanyang Technological University (NTU), Singapore. He is currently an Associate Professor at NIE, and his research projects include physiological responses to exercise and adaptations to health and sports performance. In addition, his interests are in investigating the influence of genetic factors on exercise-related outcomes and is currently involved in examining physiological predictors of human performance.

Dr Bala has extensive teaching experience in topics that include anatomy and physiology; human physiology; human functional anatomy; applied exercise physiology; physiological bases of exercise; measurement and evaluation; metabolic and cardio-respiratory aspects of exercise;

nutrition; obesity; health and fitness; hockey; track and field; and fitness and conditioning. He is a certified American College of Sports Medicine Health Fitness Director and a Fellow of the American College of Sports Medicine, and sits on a number of international boards and holds important positions in Asian and global societies such as the Asian Association of Sports Management; Asian Society of Kinesiology; Asian Society of Young Children; Asian Council for Health, Physical Activity & Fitness; The Foundation for Global Community Health; and Federation International D'Education Physique. He is also the current President of the Asian Council of Exercise and Sport Science. He was also the Secretary General of the International Conference for Physical Education & Sports Science (ICPESS) 2010 in conjunction with the 1st Youth Olympics Games organised in Singapore. Furthermore, he was the Executive President of the Asian Society of Young Children as he organised the conference in Singapore. In addition, he is an elected member of the prestigious Sigma Xi, and a member of the European College of Sports Science, The Obesity Society, and American Physiological Society. Moreover, he also conducted numerous workshops for the American College of Sports Medicine certification in health and fitness in Singapore. He also organised the prestigious International Association of Athletics Federations (IAAF) (currently known as World Athletics)-NIE/NTU Chief Coach Youth Academy in Singapore for elite coaches from all around the world. With many first-rate published research papers, proceeding papers, books, and book chapters in the area of sport science, Dr Bala serves as the editorial board member for a number of international journals and has been invited as keynote speaker/invited speaker/presenter for various renowned conferences. He is also currently involved in implementing HOPSports® Inc. Brain Breaks® in Singapore schools and engaging in collaborative global research on Brain Breaks® for children to enjoy physical activity during classroom and physical education lessons. He has won the NTU Nanyang Award (School) and several NIE Commendation of Teaching awards. The NTU Healthy Lifestyle Committee, chaired by him, won the Singapore HEALTH Award (Platinum and Gold) given by the Singapore Health Promotion Board. NTU is the only university that was awarded the Platinum award among all corporate organisations. He also stays on campus as a faculty of residence since 2016 to interact and help local and foreign students to enjoy and live a holistic campus life. He is one of the faculty who helps run "Spartans", which is an exercise programme catering to in-house student residents. It has become one of the favourite programmes for student residents and is also expanded and available online to cater to more students, allowing them to enjoy exercising and to take a breather from their hectic campus life. He also volunteered as a committee member in the organisation of the Ministry of Home Affairs (Police) Real Run for 15 years, the Ministry of Education Olive Run for 12 years and counting, and the Kebun Bahru Link Resident Committee 5-km run for many years till 2015. These running events were organised for civil servants and the public, which started off as a small-scale event but eventually became grand and popular.

From a school boy athlete (hockey, track and field, and cross-country) to a performance athlete (track and field, road races, and cross-country) who had represented his country (Singapore) in long-distance running events, he had won medals in various international and local meets. He had also earned the distinction of having qualified and raced in the prestigious National Collegiate Athletic Association (NCAA) cross-country championships in the USA. Dr Bala holds both Level I and Level II IAAF (currently known as World Athletics) coaching certificates. A team manager at the 2009 and 2015 Southeast Asian (SEA) Games, as well as the 2016 Youth SEA Games, he has coached many national and local schools' long-distance athletes who went on to achieve national honours and established national and local school records. He also volunteered and was the Assistant Honorary Secretary (2014–2016 June) and Vice President of training and selection (2016–2018 November) for the national Singapore Athletics Association. Dr Bala currently provides voluntary services to Singapore athletes as their national coach. He is currently the President of Cougars Athletic Association, an associate affiliate with Singapore Athletics, which is a non-profit club that nurtures young athletes for the future. Dr Bala has vast experience in the area of athletics coaching and is interested in vast aspects surrounding performance in track and field including sports science in various sports.

Co-Authors

Dr Visvasuresh Victor Govindaswamy

Dr Victor graduated with a Master of Science in Computer Science and Engineering and a Bachelor of Science in Electrical and Computer Engineering from the University of Texas in Austin, USA. He has a PhD from the University of Texas in Arlington and is currently an Associate Professor, Director of Computer Science Programs at Concordia University, Chicago, USA. He has worked on a number of research projects, a notable one being in the area of preparing dependable, dynamic real-time application systems for an adaptive resource management environment, which was sponsored by the Defense Advanced Research Projects Agency (DARPA), USA, the central research and development organisation for the Department of Defense and the National Aeronautics and Space Administration (NASA). Dr Victor has a strong interest in physiological research and has collaborated on a number of research projects. He is also an avid runner and has participated in many races during his school and university days.

Jolene Lim

Jolene graduated from Loughborough University, United Kingdom, with a Master of Science in Sport and Exercise Science and worked as a research assistant at the Nanyang Technological University, Singapore. She is currently pursuing her PhD in Sports Science at the Nanyang Technological University and presently working at Singapore Shooting Association. She volunteers as Vice President Development at Cougars Athletic Association. She also deals with research associated with applied sports science in children, youth, and adults, and is keenly involved in exploring the physiological aspects of human performance. She is also a sports enthusiast and engages in regular physical activities.

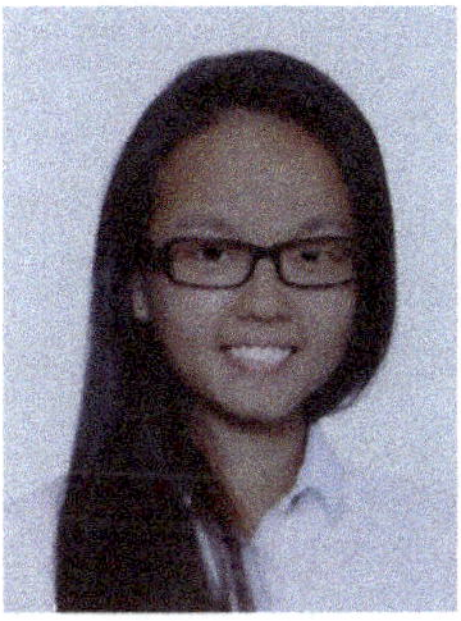

Peggy Boey

Peggy is a Physical Education teacher and is currently pursuing her Master of Science from the National Institute of Education, Nanyang Technological University. She graduated with a Bachelor in Physical Education & Sports Science at the National Institute of Education. She is a member of the American College of Sports Medicine and a life member of the Asian Council of Exercise & Sports Science. She has been involved in intensive research on children exercising within a safe intensity using the OMNI Rate of Perceived Exertion Scale and has implemented it in her teaching of Physical Education at her school. Currently, she is involved in human performance research. She also volunteers as Vice President Competitions Organising at Cougars Athletic Association. Peggy is a former competitive swimmer and has competed in long-distance races in Singapore. She also holds the Level I International Association of Athletics Federations (IAAF) coaching certificate and is looking forward to coaching younger athletes in track and field.

Ng Yew Cheo

Yew Cheo is a national athlete who obtained her Diploma in Physical Education & Sports Science at the National Institute of Education, and is a degree holder in Exercise Science from the Singapore University of Social Sciences. She is also a member of the American College of Sports Medicine and a life member of the Asian Council of Exercise & Sports Science. Furthermore, she is a Future Leader Volunteer of The Foundation for the Global Community Health, promoting physical activity in the community. She is involved in assisting the implementation of HOPSports® Inc. Brain Breaks® in Singapore schools and engaging in collaborative global research on Brain Breaks® for children to enjoy physical activity during classroom and physical education lessons. She is a former Physical Education teacher and is actively involved in track and field and trains younger athletes to reach their potential. She has represented the nation and won medals at international and national meets and is currently running competitively. She also coaches voluntarily and volunteers as Honorary Secretary at Cougars Athletic Association. On top of that, she works on research projects engaging children and adolescents in using the OMNI Rate of Perceived Exertion Scale during exercise and Physical Education lessons. Currently, she is involved in human performance research. Additionally, she holds the Level I International Association of Athletics Federations (IAAF) coaching certificate.

Acknowledgements

The main author, Dr G. Balasekaran, would like to give special thanks to Professor Robert J. Robertson (PhD, Physiology of Exercise) and Professor Robert E. Ferrell (PhD, Molecular Genetics) from the University of Pittsburgh for their utmost, invaluable advice and guidance during his PhD and post-doctoral academic journey, respectively, which eventually led to the development of this applied physiology textbook. Additionally, he would also like to specially thank Dr Silva Arslanian, MD (Endocrinologist), from the Children's Hospital of Pittsburgh, whom he worked with on research projects during his doctoral days. Many thanks to his professors, Professor James G. Mill, Professor Elaine Blair, Professor Archie Moore, and Professor Edward Sloniger of Indiana University of Pennsylvania, Department of Kinesiology, Health, and Sport Science for their mentorship, guidance, and support, and teaching their invaluable knowledge during his masters and undergraduate days. A special thanks to Professor Stephen L. Kopecky, Cardiovascular Medicine, Mayo Clinic and Professor Theodore J. Angelopoulos, Chair of the Department of Rehabilitation and Movement Sciences, University of Vermont, for writing the Foreword for the book.

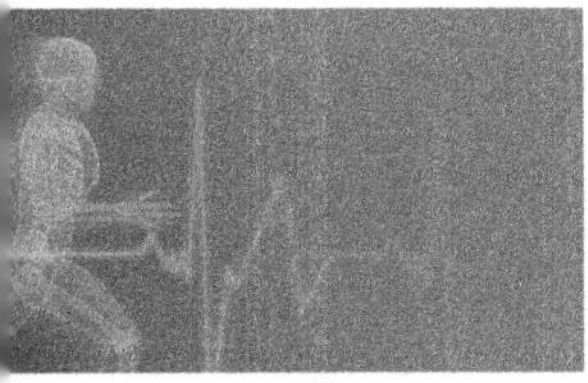

Contents

Introduction

What is Exercise Physiology?

"Exercise Physiology has evolved from its parent discipline, anatomy and physiology. Anatomy is the study of an organism's structure, or morphology. From anatomy, we learn the basic structure of various body parts and their interrelationships. Physiology is the study of body function. In physiology, we study how our organ system, tissues, cells, and molecules within cells work and how their functions are integrated to regulate our internal environments. Because physiology focuses on the functions of structures, we can't easily discuss physiology without understanding anatomy. Exercise physiology is concerned with the study of how the body adapts physiologically to the acute stress of exercise, or physical activity, and the chronic stress of physical training. Sport Physiology grew out of exercise physiology. It applies exercise physiology to problems unique to sport."

David Costill, an eminent physiologist

Terminology

Glycogenesis is the process by which glycogen is synthesised from glucose and stored in the liver and muscle.

Glycogenolysis is the process by which glycogen is broken down to glucose-1-phosphate. The enzyme which catalyses this process is called phosphorylase (Chapter 3 Table 1 & Figure 1).

Heat Production

60–70% of the energy used by the human body is released as heat and the remaining energy is used for mechanical work and cellular activities (Wilmore & Costill, 1994).

Nutrients

The human body derives its energy from carbohydrates, fats, and proteins. Carbohydrates are stored in the muscles and liver and are limited to less than 2,000 kcal of energy. This is equivalent to the energy needed to run 32 km (20 miles). On the other hand, fat stores usually surpass 70,000 kcal of energy (Wilmore & Costill, 1994).

$$1 \text{ g of carbohydrates} = 4 \text{ kcal of energy}$$
$$1 \text{ g of fat} = 9 \text{ kcal of energy}$$

(Powers & Howley, 2009)

Even though 1 g of fat can generate 2.25 times more energy than carbohydrates, more oxygen is needed to metabolise fat. Carbohydrates energy is more accessible compared to protein and fat.

Energy for Physical Activity

Continual energy supply is necessary for human activity. Energy is mainly provided through the metabolic degradation of carbohydrates and fats.

Carbohydrates are metabolised through glycolysis and the Krebs cycle. Fats are also metabolised through glycolysis but begin with a process called beta oxidation (Chapter 4 Figure 6).

Nutrient Energy Metabolism

Nutrient energy metabolism begins with carbohydrates. It is the preferred fuel of muscles and is the only nutrient that can be used to generate adenosine triphosphate (ATP) anaerobically (without oxygen).

During light and moderate exercises, carbohydrates supply approximately one-half of the body's energy requirement. However, during vigorous exercises, stored glycogen and blood glucose must predominantly supply energy needed for ATP resynthesis (Gupta & Balasekaran, 2013).

A continual breakdown of some carbohydrates is required so that fat can be used for energy.

Adenosine Triphosphate (ATP)

Energy derived from food is stored in an energy-rich compound called adenosine triphosphate. More commonly known as ATP, it is the energy currency of the cell and is needed for every operation in the cell that requires energy. Without ATP, there will be no work performed.

At any moment, there is only a limited quantity of ATP in a muscle cell. ATP is constantly being used and regenerated, and the regeneration of ATP requires energy.

ATP can be produced via 3 energy systems:

1) ATP-PC system
   Energy for ATP resynthesis comes from only 1 compound known as phosphocreatine (PCr) (Chapter 2 Figures 1 & 2).
2) Anaerobic glycolysis or lactic acid system
   Energy for ATP resynthesis comes from the partial degradation of glucose or glycogen (Chapter 2 Figure 3, Chapter 3 Figure 2, Chapter 4 Figure 6).

3) Oxygen system or aerobic glycolysis

Energy for ATP resynthesis comes from the completion of carbohydrates and fatty acids oxidation. The final routes of oxidation are the Krebs cycle and electron transport system (Chapter 2 Figure 3, Chapter 4 Figure 6).

Summary of Anaerobic Energy Systems That Produce ATP

1) ATP-PC system

Also known as ATP-CP system, ATP-PCr system, phosphagen system, phosphocreatine system (Chapter 2 Figures 1 & 2).

The main function of the ATP-PC system is to maintain ATP levels. The enzyme creatine kinase works on phosphocreatine to separate inorganic phosphate (Pi) from creatine. Pi is then combined with adenosine diphosphate (ADP) to form ATP.

1 mole of PCr = 1 mole of ATP

(Wilmore, Costill, & Kenney, 1994)

2) Anaerobic glycolysis

Also known as the lactic acid system.

In this system, either glucose or glycogen is broken down to pyruvate (pyruvic acid) by a series of glycolytic enzymes. This process is known as glycolysis. Without the presence of oxygen, pyruvate is converted to lactate (lactic acid) (Chapter 3 Figures 1 & 2).

1 mole of glucose = 2 moles of ATP

1 mole of glycogen = 3 moles of ATP

(Wilmore, Costill, & Kenney, 1994)

Use of Anaerobic Systems

The ATP-PC and glycolytic systems are the major energy contributors of energy in the first few minutes of high-intensity exercises.

The ATP-PC system supplies energy for activities lasting 3–15 seconds or less and supplies at least 8% of energy for maximal activities up to 2 minutes. Other authors have stated that ATP-PC dominates the energy contribution in sprint running for 4–6 seconds (Janssen, 1994) and maximally up to approximately 10 seconds (Spriet, 1995). Brown, Miller, and Eason (2006) stated that the phosphogen system is the predominant energy system for maximal activities lasting 15 seconds or less. Thus taking it as a range of 3–15 seconds is accurate.

The anaerobic metabolism (ATP-PC + lactic acid) supplies energy for activities lasting less than 2 minutes. At least 15% of anaerobic systems are used in activities that last as long as 10 minutes.

Depending on the duration of exercise, the lactic acid system becomes more important as exercise duration increases. By 5 minutes of exercise, the oxygen system (aerobic system or aerobic glycolysis) becomes the dominant system.

Anaerobic: without oxygen
Aerobic: with oxygen

Anaerobic: 2 kinds — ATP-PC system (alactic system: no generation of lactic acid) and anaerobic glycolysis system (lactic acid system: generation of lactic acid)

(Adapted from Brown, Miller, & Eason, 2006)

Substrate Utilisation During Rest

Fats: 2/3 of food fuel
Carbohydrates: 1/3 of food fuel
Protein: negligible contribution as a food fuel

Oxygen consumption during rest is 0.20–0.35 litre/minute (Wilmore & Costill, 1994). During rest, the aerobic system is the only energy system in operation and fat is the major source, as opposed to maximal exercise, where carbohydrate is the major source (Andreacci et al., 2004).

During rest, the oxygen transport system (heart and lungs) is capable of:

- Supplying each cell with sufficient oxygen
- Supplying each cell with adequate ATP
- Satisfying all the energy requirements of the resting state
- Supplying a small but constant amount of lactic acid present in the blood (10 mg/100 ml). The lactic acid level remains constant and does not accumulate (anaerobic glycolysis is not operating).

Energy Contribution by the Energy Systems During Exercise

Both anaerobic and aerobic systems contribute to ATP production during exercise. Energy systems do not "switch on" or "switch off". One energy system usually predominates in an activity.

Relative roles of the energy systems are dependent on:

- Exercise type
- State of training
- Diet of the athlete
- Intensity: performed for only short periods of time but which requires maximal effort
- Duration: performed for relatively long periods of time but which requires submaximal effort

Substrate Utilisation During Short Duration Exercise

Carbohydrates: major food fuel
Fats: minor food fuel
Protein: negligible contribution as food fuel

ATP required for high-intensity exercise cannot be supplied by the aerobic system alone. Most of the ATP must be supplied **anaerobically** by the phosphagen system and anaerobic glycolysis (Chapters 2 & 3).

- PCr drops to very low levels and remains low until exercise stops. It is rapidly replenished (within minutes) during recovery.
- Majority of PCr stores are replenished by the aerobic system during recovery (Chapter 2 Figure 3).
- If exercise continues for more than 15 seconds (till 2 minutes), lactic acid will accumulate. Lactic acid values drop when exercise decreases in intensity or ceases (Chapter 3 Figure 2).

Substrate Utilisation During Prolonged Exercise

Carbohydrates: major food fuel

Fats: minor food fuel

Blood lactic acid increases but not to maximum.

For exercises longer than an hour, glycogen stores significantly decrease, and fats become more important (Chapter 4).

Fat and Carbohydrate Training

1. **Can an athlete run fast using the fat stores? Explain.**

2. **How does an elite athlete run a marathon (42.195 km) if the carbohydrate stores run out at approximately 32 km?**

3. **Does the novice runner have carbohydrate stores which can last for approximately 32 km? Explain.**

4. **Briefly describe the intensity and duration concept in relation to substrate utilisation. Does this concept of substrate utilisation based on intensity and duration differ for elite athletes?**

5. **What distance should a person run to burn 1 pound of fat on a standardised 400-m track (approximately 100 kcals energy expenditure for 1.6 km) if he/she burns 500 kcals per exercise session? What distance should he/she run a day on the track, and for how many days, in order to burn 1 pound of fat? (1 pound = 3,500 kcals) (this calculation is without food intake, only expenditure)**

Chapter

2

ATP-PC System

The ATP-PC system is where adenosine triphosphate (ATP) is most readily available for muscles to use as it does not require a long series of chemical reactions or the transportation of oxygen to the working muscles.

ATP and phosphocreatine (PCr) are stored in the contractile mechanisms of the muscle cells and both contain phosphate groups. Only a small amount of ATP is stored within the cell. A large amount of energy is released when the phosphate group is removed.

Phosphate is removed with the help of enzymes, which are protein compounds that accelerate the speed of individual reactions (adenosine triphosphatase (ATPase) for ATP, creatine kinase for PCr) (Figures 1 & 2).

ATP is continuously recycled in each cell and cannot be supplied by blood or other tissues. In order to sustain it during exercise, ATP must be synthesised at the rate it is being used.

Utilisation of ATP-PC System

ATP-PC is needed for maximal short-duration activities that require immediate and rapid energy supply such as a 50-m dash, 100-m dash (usually elite athletes), 25-m swim, weightlifting, high jump, long jump, discus, javelin throw, hammer throw, basketball lay up, and volleyball spike.

ATP-PC stored within the activated muscles during the specific exercises provides energy needed to perform such short-duration, high-intensity activities.

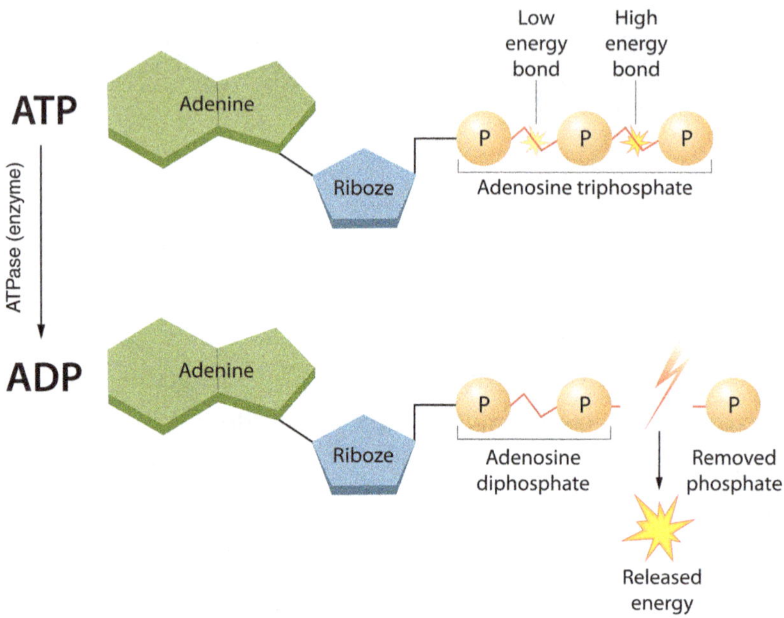

Figure 1. Energy released from ATP for human performance.

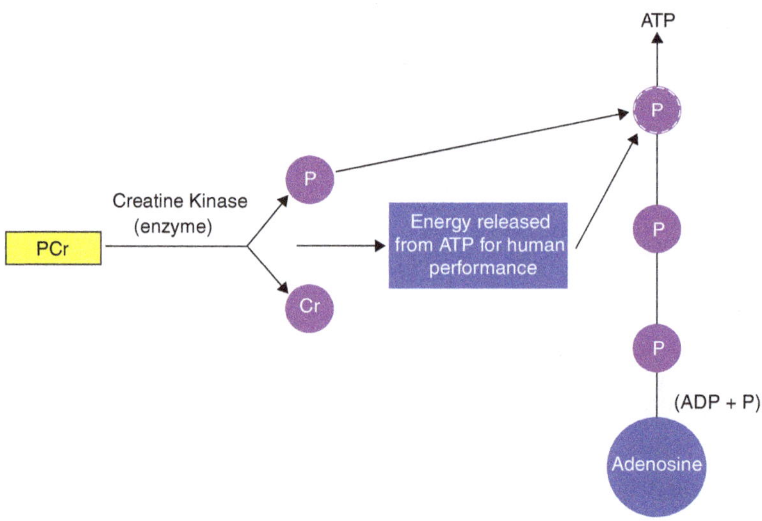

Figure 2. Splitting of creatine phosphate (PCr) produces energy and phosphate (P), which are used to resynthesise ATP from ADP and P.

Recovery of ATP-PC System

The reformation of PCr from inorganic phosphate (Pi) and creatine is only possible from the energy released during the breakdown of ATP.

The aerobic system predominantly replenishes PCr stores immediately after exercise, with a small contribution from anaerobic glycolysis (Figure 3). If the intensity of exercise is very high, the immediate replenishment of PCr stores comes from anaerobic glycolysis followed by aerobic glycolysis as the intensity decreases slowly. However, if the intensity of exercise is low, the immediate replenishment can come from aerobic glycolysis (Figure 3). At longer recovery periods (after exercise has completely terminated the primary source of ATP), PCr stores are replenished through the breakdown of food. The stored carbohydrates, fats, and proteins are ready to continually recharge the phosphate pool. Thus, after PCr stores

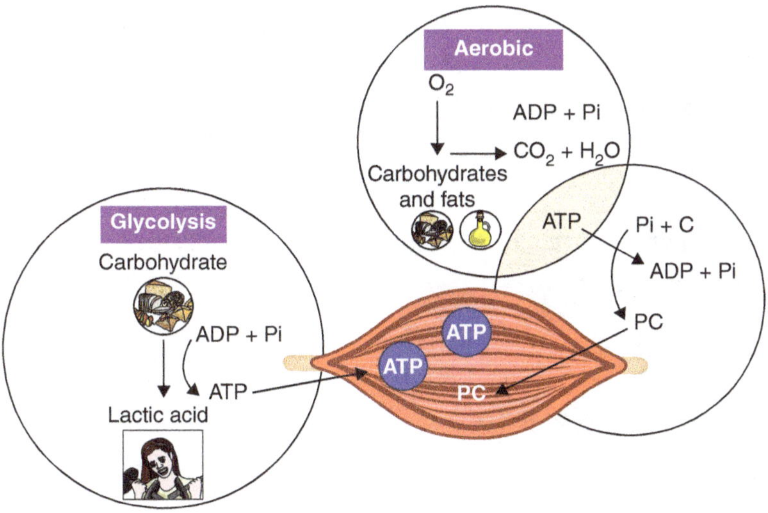

Figure 3. Oxygen consumed during the fast component of recovery provides the majority of energy necessary to replenish ATP and phosphocreatine (PC) stores in muscles that were depleted during exercise. Some of the ATP resynthesised is directly stored in muscles and some break down immediately to resynthesize PC, which is then stored in the muscles. Anaerobic glycolysis may also provide some energy (ATP) for phosphagen restoration (Bowers and Fox, 1993). This usually happens if exercise intensity is very high and the oxygen system may be too slow to replenish the stores.

are depleted from high-intensity exercises, they cannot be effectively replenished until recovery has started.

In order to sustain or recover from an all-out effort exercise, additional energy needs to be generated to replenish ATP. For example, after a high-intensity workout, the athlete must be able to readily replenish PCr stores or he/she would not be able to maximally perform the next bout of exercise. Thus, you need a well-trained aerobic and anaerobic glycolysis system in order to replenish PCr stores.

ATP-PC Training

1. **Outline examples of an ATP-PC workout. Explain why.**
2. **How long does it take for 70% of ATP-PC stores to be replenished? How long do we allow students in schools to rest in between repetitions for an ATP-PC workout?**
3. **Based on research, how long does it take for PC stores to be fully replenished?**
4. **Give examples of a structured ATP-PC workout.**
5. **Give examples of an unstructured ATP-PC workout.**

6. **Give examples of structured ATP-PC workouts where full rest in between repetitions are given.**

7. **Is a 30-second Wingate maximal cycling test a good test to evaluate the ATP-PC system? Explain with a physiological rationale (refer to *Applied Physiology of Exercise Laboratory Manual* Laboratory Session 10, Balasekaran, Govindaswamy, Lim, Boey, & Ng, 2021).**

Summary of the ATP-PC System

1. **In this system, energy needed for adenosine triphosphate (ATP) resynthesis comes from only 1 compound: phosphocreatine (PCr).**
2. **PCr in muscle cells have about 4–6 times more concentration than that of ATP (McArdle, Katch, & Katch, 2007, 2010). Thus, PCr is considered the high energy phosphate "reservoir".**
3. **The system does not depend on**
   - **a long series of chemical reactions**
   - **the transportation of oxygen to the working muscles**
4. **Both ATP and PCr are stored directly within the contractile mechanisms of the muscle.**
5. **The phosphagen system represents the most rapidly available source of ATP for use by the muscle.**

Chapter 3

Anaerobic Glycolysis

Anaerobic glycolysis involves the incomplete breakdown of carbohydrates to lactic acid without the presence of oxygen. Carbohydrates are converted to glucose (or stored as glycogen in the liver and muscle) and undergo a series of reactions collectively termed as "glycolysis" (which literally means "the splitting of glucose").

Stored glycogen must first be converted to glucose through a process called glycogenolysis. This breaks glycogen down to glucose-1-phosphate. Glucose and glycogen must first be converted to glucose-6-phosphate for glycolysis to begin.

(Foss & Keteyian, 1998)

Glycolysis occurs in the cytoplasm (watery medium of the cell outside the mitochondria).

Table 1: Important reactions and their key enzymes.

Reactions	**Key Enzymes**	**Enzymes' Abbreviations**
Glucose to Glucose-6-phosphate	Hexokinase	HK
Fructose-6-phosphate to Fructose 1,6-bisphosphate	Phosphofructokinase	PFK
Phosphoenolpyruvate to Pyruvate	Pyruvate Kinase	PK
Pyruvate to Lactate	Lactate Dehydrogenase	LDH
Glycogen to Glucose-1-phosphate	Phosphorylase	–

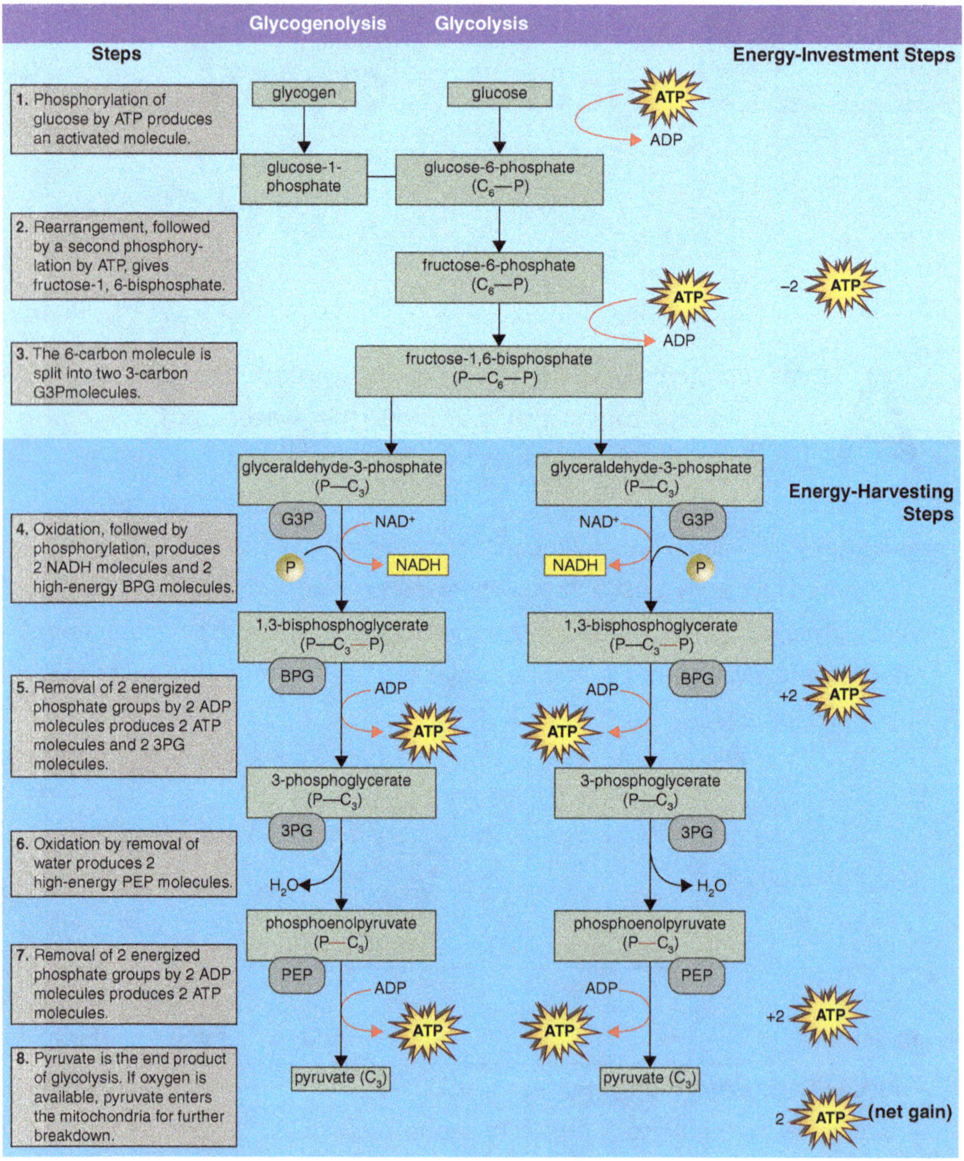

Figure 1. Glycolysis occurs in the cytoplasm (adapted from Powers & Howley, 2009).

The first 3 reactions mentioned are irreversible reactions, and the enzymes associated with these reactions are rate-limiting enzymes (Table 1, Figure 2).

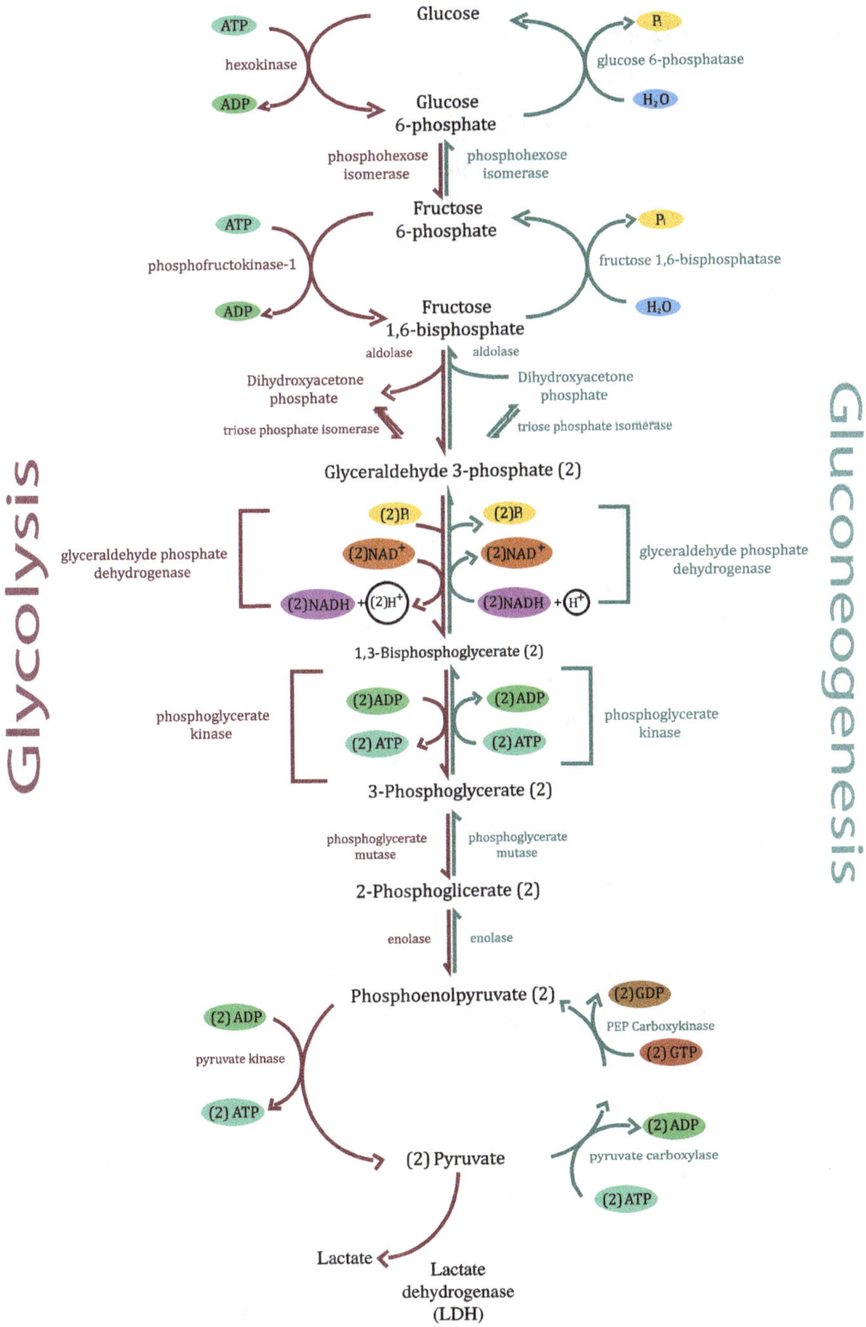

Figure 2. The glycolysis process with their key enzymes.

Rate-Limiting Enzymes

Control and limit the overall performance of glycolysis.

The glycolysis process might be impeded if these rate-limiting steps are slow or insufficient. Hence, coaches should focus on building up more of these enzymes to speed up the process of glycolysis (Figure 2).

Lactate

Blood lactate concentration represents the systemic balance between net lactate production and clearance (Balasekaran, 2002; Goodwin et al., 2007). It is the end product of anaerobic glycolysis (Figure 2).

A. Exercise Intensity and Lactate

During high-intensity or long-duration exercises (where fatigue occurs), lactate accumulates in the blood (lactate production > lactate clearance). High lactate accumulation during exercise will cause fatigue and reduce the pace during a run or cease physical activity. An unfit individual can accumulate lactate even during a short duration of exercise since his body is not aerobically fit enough to delay the onset of fatigue (lactate).

The increase in acidity in muscles (due to the increase in hydrogen ions) reduces exercise intensity as increased acidity interferes with calcium and contractile elements in the muscles, as well as obstructs PFK activity, which is important in anaerobic glycolysis (Foss & Keteyian, 1998).

Increased acidity also interferes with muscle glycogen breakdown. In longer-duration races, there is a diminished capacity to mobilise lipids. At the same time, glycogen breakdown is slowed, leading to a reduction in endurance time (Plowman & Smith, 2003).

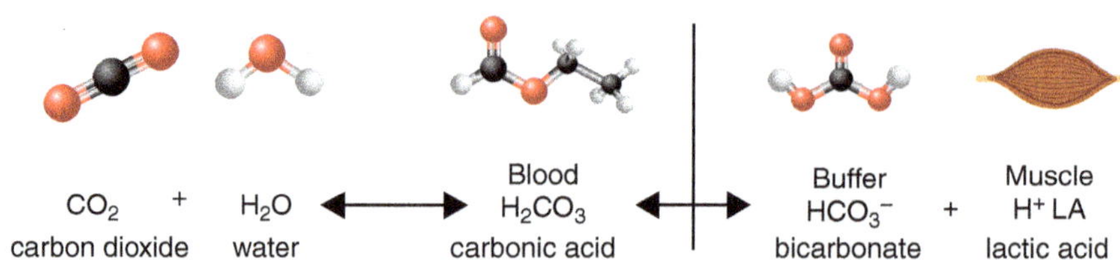

Figure 3. Pulmonary ventilation removes H^+ from blood by the HCO_3^- reaction (Foss & Keteyian, 1998).

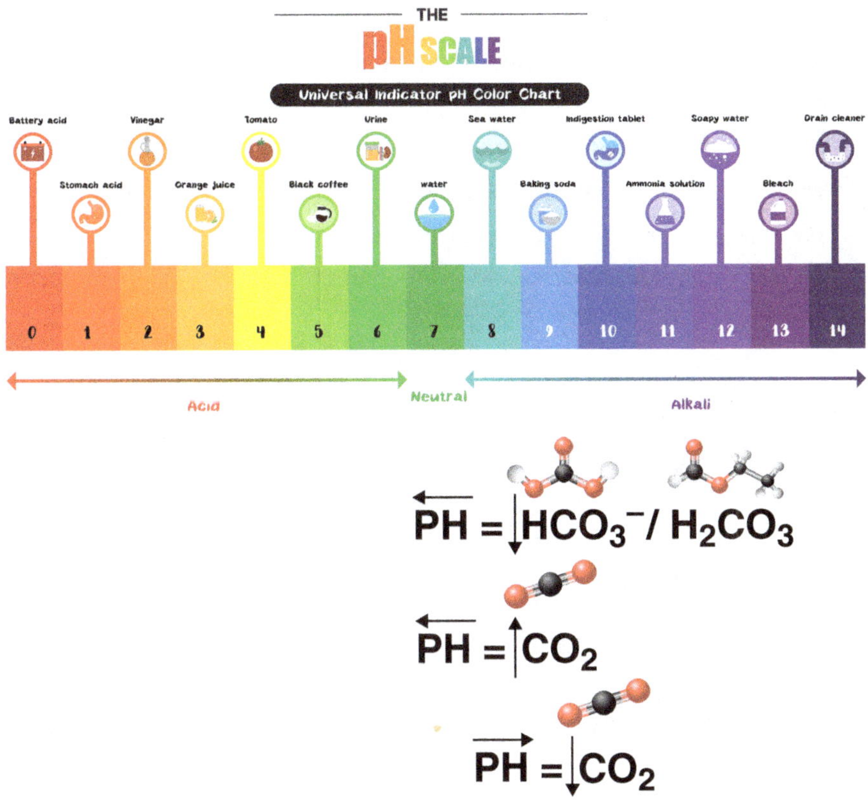

Figure 4. pH values are affected by pulmonary ventilation (adapted from Foss & Keteyian, 1998).

B. Removal of Lactate

Increased ventilation results in CO_2 exhalation: H^+ concentration decreases → pH increases (Figures 3 & 4)

Decreased ventilation results in build-up of CO_2: H^+ concentration increases → pH decreases (Figures 3 & 4)

C. Reasons for Fatigue

According to Holloszy (1982), there appears to be a maximum amount of lactic acid that can accumulate before an individual stops exercising due to severe muscular fatigue.

This could be due to a drop in intracellular pH as a result of lactic acid accumulation, inhibiting the rate-limiting PFK activity which is important in anaerobic glycolysis (Foss & Keteyian, 1998).

D. Cori Cycle and Gluconeogenesis

Lactate or pyruvate produced by anaerobic glycolysis is eventually cleared through the Cori cycle (Figure 5). Glucose produced in this way is channelled back to glycolysis to be used as an energy source in performing work. This pathway can also be used by reducing your exercise intensity if you have produced a high level of lactate via your high-intensity workout, which affects your muscle contractile elements and substrate utilisation

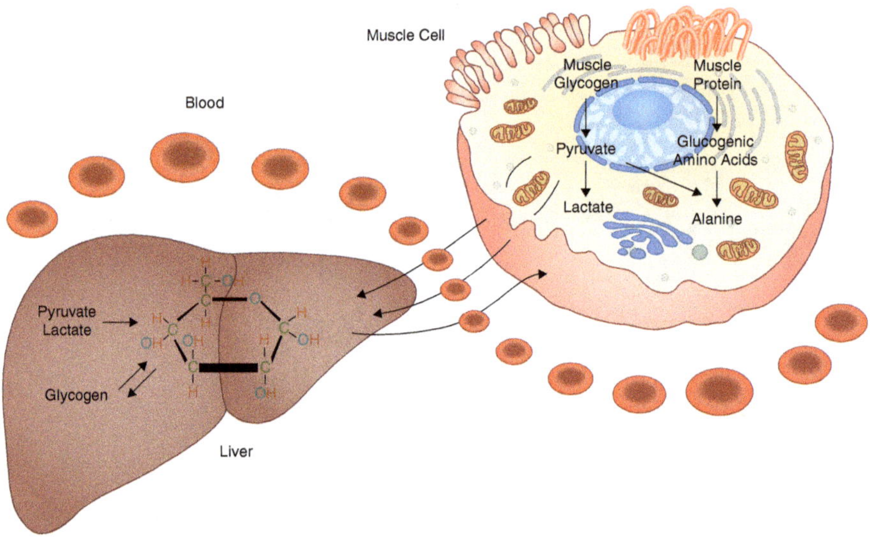

Figure 5. The Cori cycle provides for the removal of lactic acid formed in the muscle with a resulting increase in available glucose. This process is also known as gluconeogenesis (adapted from McArdle, Katch, & Katch, 2006).

of glycogen. This re-utilisation of lactate can be trained whereby cardiovascular and physiological adaptations can take place so that it can be used tactically as an energy source in track races (more than 800-m events) or in sports/track races which require short duration, high-speed intensity via anaerobic glycoylsis and produce lactate (Figures 1 & 2).

E. Moles of ATP Produced During Anaerobic Glycolysis

Only a few moles of ATP can be resynthesised from glucose or glycogen during anaerobic glycolysis:

- 1 mole of glucose produces 2 moles of ATP
- 1 mole of glycogen produces 3 moles of ATP

(Wilmore & Costill, 1994)

Anaerobic Glycolysis Training

1. **Outline examples of an anaerobic glycolysis workout.**

2. **What is the time frame for lactate clearance after a high-intensity anaerobic glycolysis exercise/workout? Explain the implications.**

3. **Can the rest time between high-intensity lactate workouts be decreased? What kind of training workouts will bring about this adaptation to decreased rest time? Can you list the physiological rationale for this adaptation?**

4. **Give examples of a structured anaerobic glycolysis workout.**

5. **Give examples of an unstructured anaerobic glycolysis workout.**

6. **Can an athlete reuse his/her lactate as an energy source and use it to his/her advantage (Hint: Figures 1, 2, & 5)? What structured and unstructured exercise workouts would you do for such adaptations to occur?**

Summary of Anaerobic Glycolysis

1. Anaerobic glycolysis results in the formation of lactic acid, which is related to muscular fatigue.
2. Anaerobic glycolysis does not require the presence of oxygen.
3. There are important reactions in anaerobic glycolysis. Some of these reactions are rate-limiting.
4. Anaerobic glycolysis uses only carbohydrates (glycogen and glucose) as its food fuel.
5. Anaerobic glycolysis releases energy for the resynthesis of only a few moles of ATP.
6. Lactate formed by anaerobic glycolysis is eventually cleared via the Cori cycle.

Chapter 4

Aerobic Metabolism

Carbohydrate Metabolism

During aerobic glycolysis, 1 mole of glucose can be completely broken down to CO_2 + H_2O + 38 moles of adenosine triphosphate (ATP). 39 moles of ATP are generated if glycogen is used (Wilmore & Costill, 1994).

Such a large energy yield requires many reactions that can be divided into 3 main reactions:

1. Aerobic glycolysis
2. Krebs cycle/citric acid cycle
3. Electron transport chain (electron transport system)

Acetyl group is a two-carbon molecule.

1) Pyruvic acid (a three-carbon molecule) releases carbon dioxide (CO_2) during oxidation to become an acetyl group before entering the Krebs cycle (Figure 1).
2) In fatty acid metabolism, which follows the same route as carbohydrate metabolism, two-carbon groups are also formed through beta oxidation before entering the Krebs cycle.

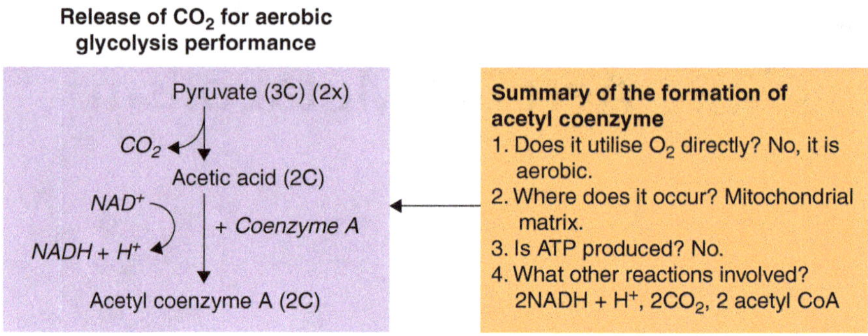

Figure 1. The formation of acetyl coenzyme (adapted from Plowman & Smith, 2003).

Oxidation

A coenzyme is a non-protein substance associated with activating an enzyme, typically a vitamin.

Two very important coenzymes of oxidative pathway:

1) Nicotinamide adenine dinucleotide: $NAD^+/NADH$
2) Flavo adenine dinucleotide: $FAD^+/FADH_2$

NAD^+ and FAD^+ are derived from B vitamins niacin and riboflavin, respectively.

During glycolysis and the Krebs cycle activity, H^+ is removed from carbohydrates. This is known as oxidation. A compound is said to have undergone reduction if it accepts H^+ ions (e.g. NAD^+ and FAD^+ undergo reduction to form NADH and $FADH_2$).

NAD^+ and FAD^+ carry and accept hydrogen ions, while NADH and $FADH_2$ carry electrons through the electron transport chain (Figures 2 & 3).

(Foss & Keteyian, 1998)

Hydrogen

Hydrogen atoms are continually removed from stored carbohydrates, fats, and proteins during energy metabolism. Electrons from hydrogen atoms are removed (oxidation) by carrier molecules in the mitochondria and eventually transported to oxygen (reduction) (McArdle, Katch, & Katch, 2010).

Oxygen also accepts hydrogen to form H_2O. The energy generated through this process is trapped as chemical energy in the form of ATP.

Krebs Cycle (Citric Acid Cycle)

2 ATP formed for every mole of glucose.

There are 4 sites in the Krebs cycle where H^+ ions are removed (Figure 2) and passed through the electron transport chain for the formation of H_2O and ATP (Figures 3 & 5).

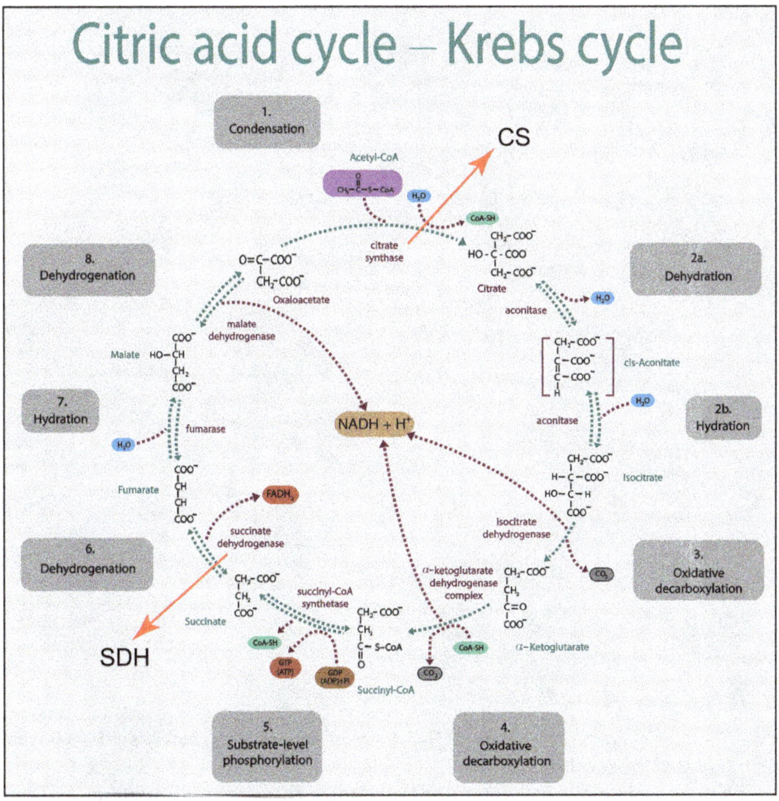

Figure 2. The Krebs cycle.

Key Enzymes (Figures 2, 3, & 5)

Stages	Key Enzymes	Abbreviations
Krebs Cycle	Citrate Synthase (reaction 1–2)	CS
	Succinate Dehydrogenase (reaction 5–6)	SDH
Electron Transport Chain	Cytochrome Oxidase	CcO/Cyt

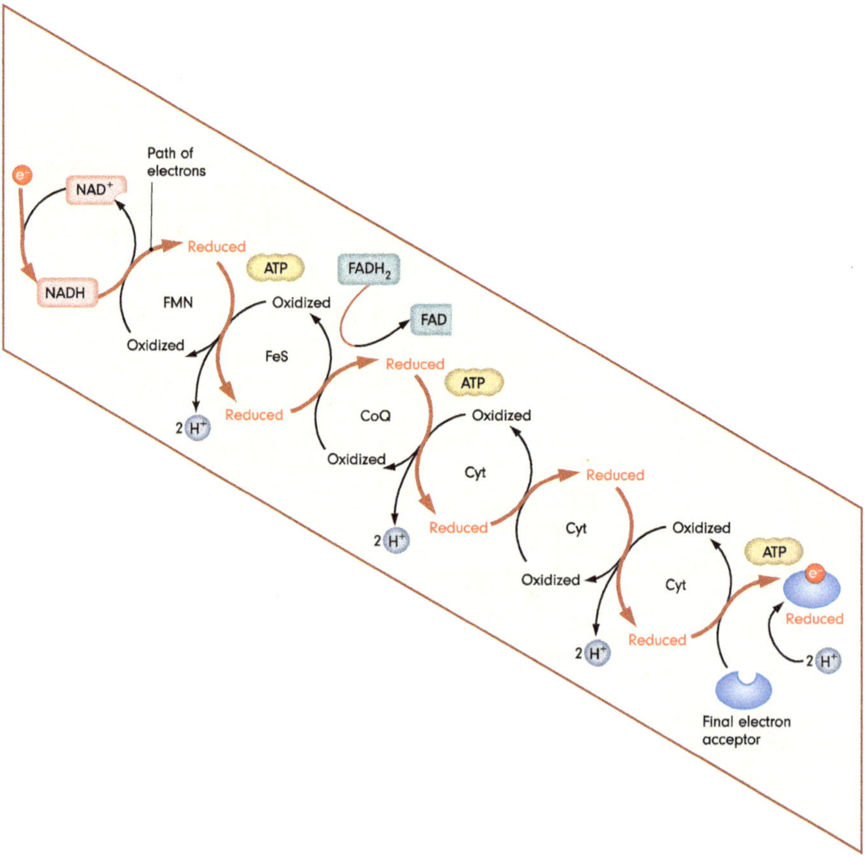

Figure 3. Electrons carried by NADH and $FADH_2$ are transported by cytochromes to form water (H_2O) at the end of the reaction.
Acronymns: NAD^+ (nicotinamide dinucleotide$^+$); NADH (nicotinamide adenine dinucleotide (NAD) + hydrogen (H)); FMN (flavin mononucleotide); FeS (non-heme iron); CoQ (coenzyme Q10/ubiquinone); $FADH_2$ (flavin adenine dinucleotide (reduced)); Cyt (cytochrome).

Electron Transport Chain (Electron Transport System)

H^+ ions from the Krebs cycle enter the electron transport chain via NADH and $FADH_2$ and are transported to molecular oxygen to form water (Figures 3 & 5).

Oxidative phosphorylation is a complicated process where ATP is synthesised during the transfer of electrons from NADH and $FADH_2$ to molecular oxygen. This process is the cell's main way of extracting and trapping chemical energy in the form of high energy phosphates (ATP).

The oxidative phosphorylation process can be likened to a waterfall separated into a series of cascades by the intervention of waterwheels at different heights, where the waterwheels harness the energy of the falling water (Figure 4). Similarly, the electrochemical energy generated in the electron transport system from one respiratory chain component to the

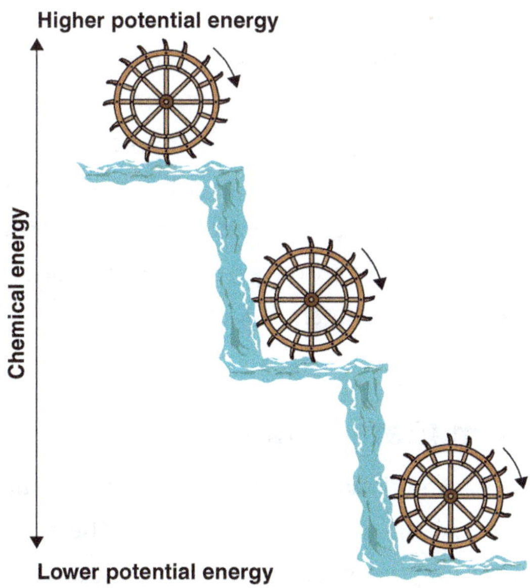

Figure 4. Oxidative phosphorylation (adapted from McArdle, Katch, & Katch, 2007).

next is harnessed and transferred or coupled to adenosine diphosphate (ADP). The energy in NADH is transferred to ADP to form ATP again at 3 distinct coupling sites during the electron transport system. This oxidation of hydrogen and subsequent phosphorylation can be summarised as follows:

$$NADH + H^+ + 3\ ADP + 3\ P + \tfrac{1}{2}\ O_2 \rightarrow NAD^+ + H_2O + 3\ ATP$$

(Adapted from McArdle, Katch, & Katch, 2007) (Figure 5)

Transfer of Electrons in the Electron Transport Chain

The electrons carried by NADH and $FADH_2$ are then passed through a system in a "bucket brigade" fashion by a series of iron protein electron carriers known as cytochromes. These cytochromes continually donate electrons to the next cytochrome through oxidation and reduction reactions.

The transport of electrons by specific carrier molecules constitutes the respiratory chain. Of the 5 cytochromes, the last cytochrome oxidase discharges its electron directly to molecular oxygen (Figures 3 & 5).

Free energy is released in the respiratory chain in relatively small amounts and in several of the electron transfers. Energy is conserved through the formation of high energy phosphate bonds (Grantham & Balasekaran, 2004; McArdle, Katch, & Katch, 2007).

Fat Metabolism (Beta Oxidation)

Fats must first be broken down to triglycerides or free fatty acids (FFAs) through lipolysis. FFAs are then transported to the mitochondria and undergo beta oxidation, a process by which successive pairs of carbon atoms from FFAs are broken off. These carbon atoms are used to form acetyl-CoA before entering the Krebs cycle and subsequently the electron transport chain (Figures 2, 3, & 5) (Plowman & Smith, 2003).

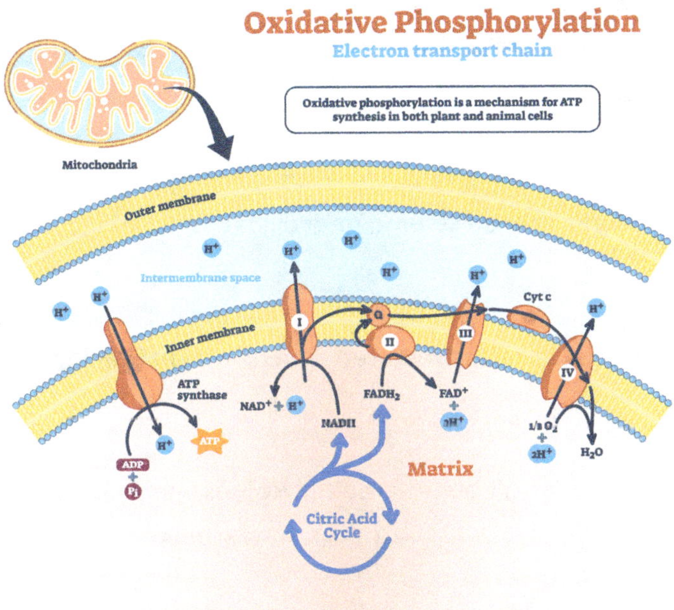

Figure 5. The electron transport chain.

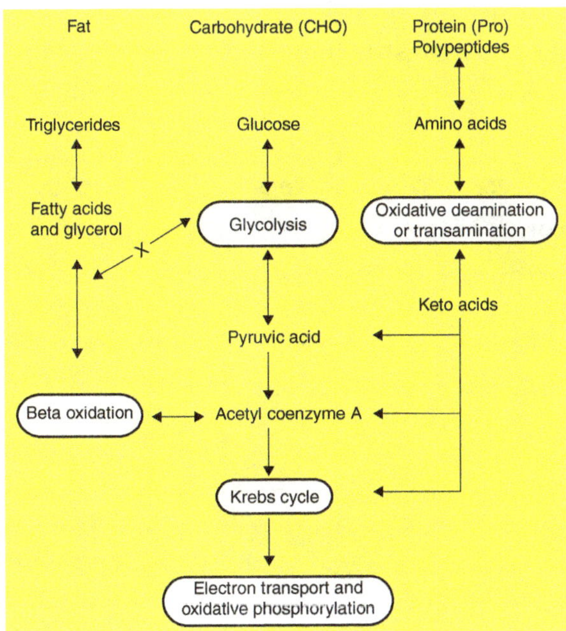

Figure 6. Cellular respiration overview (adapted from Plowman & Smith, 2003).

1. **How does one utilise the aerobic system using beta oxidation for training?**

2. **Give examples of structured beta oxidation training.**

3. **Give examples of unstructured beta oxidation training.**

4. **How does one utilise the aerobic glycolysis energy system, which uses carbohydrates as substrates, for training?**

5. **Give examples of structured aerobic glycolysis training.**

6. **Give examples of unstructured aerobic glycolysis training.**

7. **Refer to Figure 6. Why is there an "X" between fat metabolism and glycolysis? Explain.**

Oxygen Uptake

Oxygen utilisation by working muscles must be satisfied by a commensurate oxygen uptake or fatigue will quickly ensue.

Oxygen Uptake by Working Muscles

An individual's ability to transport oxygen to the working muscles is dependent on both central (cardiac output) and peripheral (arterial-venous oxygen difference) factors (adapted from Plowman & Smith, 1997) (Figure 1).

$$\textbf{Oxygen Uptake } (\mathbf{VO_2}) = \mathbf{Q} \times (\mathbf{a} - \mathbf{VO_2}\textbf{diff})$$
$$\mathbf{Q} = \mathbf{HR} \times \mathbf{SV}$$

Arteriovenous oxygen difference (a – VO_2diff): difference in oxygen concentration between arterial and venous blood; a measure of oxygen uptake by skeletal muscle.

Cardiac Output (Q): amount of blood pumped per unit of time; L•min^{-1}.

Heart Rate (HR): frequency of heart contraction per unit time; beats•min^{-1}.

Stroke Volume (SV): volume of blood pumped by the heart with each contraction; mL•contraction^{-1}.

At rest, cardiac output is approximately 5 L•min^{-1} (HR of ~70 beats•min^{-1} × 70 ml of blood•contraction^{-1} per beat). (a – VO_2diff) is approximately 5 ml of oxygen per 100 ml of blood (Plowman & Smith, 2003).

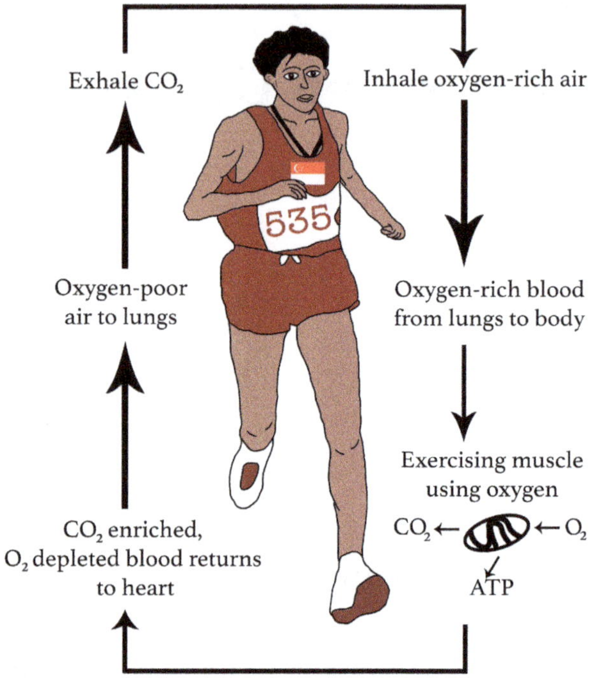

Figure 1. Oxygen utilisation during exercise.

Physiological Changes in Acute Bouts of Submaximal Exercise or Acute Dynamic Exercise

A. Oxygen

During a progressive incremental exercise test, oxygen uptake increases linearly as workload increases until VO_{2max} is reached. This increase is due to changes in cardiac output and arterial-venous oxygen difference.

$$\uparrow Q \times \uparrow (a - VO_2\text{diff}) = \uparrow \text{Oxygen Uptake}$$

$\uparrow$: increase

B. Cardiac Output (Q)

Cardiac output increases linearly as a function of oxygen uptake or workload during a progressive incremental exercise test. This is primarily due to increases in heart rate and stroke volume.

C. Heart Rate (HR)

Heart rate increases linearly as a function of oxygen uptake or workload during a progressive incremental exercise test. This is due to the withdrawal of parasympathetic tone and augmentation of sympathetic neural input to the sinoatrial node.

D. Stroke Volume (SV)

Stroke volume increases as a function of oxygen uptake or power output during submaximal exercise.

E. Arteriovenous Difference ($a - VO_2diff$)

Arteriovenous difference, which reflects oxygen uptake by skeletal muscle, increases progressively as a function of exercise intensity (5 ml of oxygen per 100 ml^{-1} to 16 ml per 100 ml^{-1}) (McArdle, Katch, & Katch, 2010) (Figure 2).

- Increases in ($a - VO_2diff$) during exercise is aided by a redistribution of blood flow due to vasoconstriction and vasodilation of blood vessels (Figure 2)
- During exercise, coronary blood flow increases in proportion to Q, but the percent of Q directed to the coronary arteries remains constant (4–5% of total Q) (Figure 2)

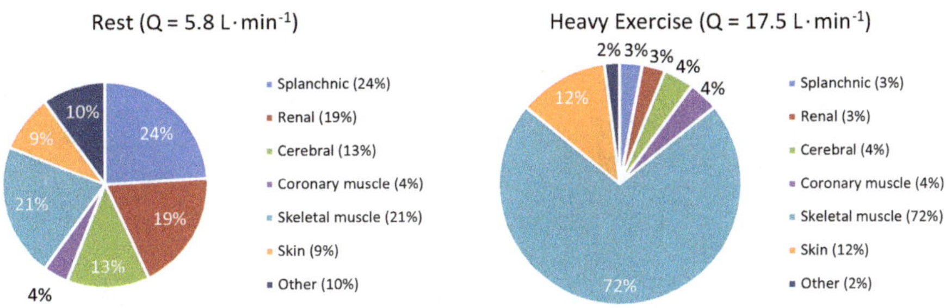

Figure 2. Distribution of cardiac output at rest and during heavy exercise (adapted from Plowman & Smith, 2003).

Maximal Dynamic Exercise

Maximal oxygen uptake (VO_{2max}) is a function of both maximal delivery and maximal extraction of circulating oxygen, as measured by the maximal cardiac output (Q_{max}) and maximal arteriovenous differences ($(a - VO_2diff)_{max}$), respectively.

$$VO_{2max} = (Q_{max}) \times (a - VO_2diff)_{max}$$

Maximal Stroke Volume

Maximal SV occurs at approximately 40–50% of maximal oxygen uptake (Plowman & Smith, 2003) due to the offsetting influences of increasing myocardial contractility and ejection fraction, as well as a progressive reduction in diastolic filling time as a result of an increase in HR.

Maximal Arteriovenous Differences $(a - VO_2diff)_{max}$

During exercise, approximately 85% of Q is directed towards the exercising muscles in comparison to the 15–20% at rest (Wilmore & Costill, 1999) (Figure 3).

Figure 3. Distribution of cardiac output at rest and during maximal exercise (adapted from Plowman & Smith, 2003).

Measurement of VO_{2max}

1. Maximal oxygen uptake (VO_{2max}): gold standard measure of functional capacity of the cardiorespiratory system.

2. VO_{2max}: highest amount of oxygen an individual can take in and utilise to produce adenosine triphosphate (ATP) aerobically while breathing air during heavy exercise.
3. Open circuit spirometry (gold standard).
4. Measure pulmonary ventilation and compare inspired and expired oxygen: oxygen consumption (VO_2: the amount of oxygen taken up, transported, and used at the cellular level) and carbon dioxide production (VCO_2: the amount of carbon dioxide generated during metabolism).
5. Treadmill and cycle ergometry are the most common modalities used to measure VO_{2max}.

There is no commonly used protocol that is satisfactory for all population of participants. Protocol is dependent on the cohort of participants that you are working with (e.g. protocols for elite athletes can differ from protocols for the healthy or special population).

Predicting VO_{2max}

VO_{2max} can be predicted through other tests such as submaximal bench stepping, running, and cycle ergometer tests. Results from these tests are highly correlated with laboratory determined VO_{2max}.

When undertaken, these tests:

- Evaluate cardiorespiratory functional capacity
- Evaluate responses to fitness conditioning
- Increase individual motivation for entering and adhering to exercise programmes
- Determine lactate threshold at percentage of VO_{2max} with lactate determination at submaximal protocol

Incremental to maximal exercise:

- Oxygen consumption during exercise of increasing intensity up to VO_{2max}
- VO_{2max} occurs in a region where a further increase in work is not accompanied by an additional increase in oxygen consumption

1. **Give examples of VO_{2max} values for different sports.**

2. **What is the highest VO_{2max} recorded?**

3. **What kinds of modality do we use to obtain VO_{2max}?**

4. **What modality was the highest VO_{2max} recorded?**

5. **How would you, as a physical education teacher/coach, conduct training to induce cardiovascular and peripheral adaptations using the formula below? Explain.**
   Oxygen Uptake (VO_2) $= Q \times (a - VO_2diff)$, where $Q = HR \times SV$.

6. **The following is an example of an individual with a VO_{2max} of 45.32 $mL \cdot kg^{-1} \cdot min^{-1}$ or 2.42 $L \cdot min^{-1}$.**

   Gender: Female — **Ambient Temperature: 18°C**
   Age: 22 yr — **Barometric Pressure: 752 mmHg**
   Weight: 53.4 kg — **Relative Humidity: 5%**

Table 1. Aerobic metabolic responses at rest and during submaximal exercise (adapted from Plowman & Smith, 2003).

Time (min)	V_ESTPD ($L{\cdot}min^{-1}$)	O_2%	CO_2%	VO_2 ($L{\cdot}min^{-1}$)	VCO_2 ($L{\cdot}min^{-1}$)	VO_2 ($mL{\cdot}kg^{-1}{\cdot}min^{-1}$)	RER	HR ($beats{\cdot}min^{-1}$)
				REST (STANDING)				
1	7.57	17.11	3.26	0.30	0.23	5.61	0.81	75
2	7.58	17.21	3.20	0.29	0.23	5.43	0.82	75
3	8.64	16.95	3.37	0.34	0.28	6.55	0.80	75
X	7.93	17.09	3.28	0.31	0.25	5.86	0.81	75
				3.5 $mi{\cdot}hr^{-1}$ WALKING (STEADY STATE); 7% GRADE				
1	21.96	15.85	4.34	1.14	0.93	21.53	0.81	120
2	24.20	15.85	4.60	1.25	1.10	23.59	0.87	136
3	23.46	15.73	4.76	1.25	1.10	23.40	0.88	136
4	26.07	16.02	4.66	1.29	1.20	24.15	0.93	136
5	25.71	15.95	4.73	1.29	1.20	24.34	0.92	136
6	27.53	16.03	4.68	1.35	1.27	25.46	0.93	136
7	25.71	15.91	4.79	1.29	1.22	24.34	0.93	136
8	28.64	16.06	4.68	1.39	1.33	26.21	0.94	136
X	25.41	15.93	4.66	1.28	1.17	24.13	0.90	136
				3.5 $mi{\cdot}hr^{-1}$ WALKING (STEADY STATE); 7% GRADE				
33	28.64	16.26	4.56	1.33	1.29	25.09	0.96	140
34	31.21	16.33	4.55	1.43	1.41	26.96	0.98	144
35	31.21	16.32	4.54	1.43	1.39	26.96	0.97	144
36	30.84	16.16	4.67	1.47	1.43	27.71	0.96	146
37	32.31	16.19	4.68	1.52	1.50	28.65	0.97	159

(a) The participant did an incremental graded exercise test (Table 1). Explain if she attained VO_{2max} based on the criteria by the American College of Sports Medicine (ACSM) guidelines (Hint: see VO_{2max} criteria on pages 39 & 40).

(b) What is her percentage of VO_{2max} attained in this test? Is this a maximal test? Describe the energy sources she used during the exercise test (Table 1).

Expressing Submaximal and Maximal Oxygen Uptake

Absolute oxygen uptake is expressed in $L{\cdot}min^{-1}$. It reflects the ability to perform external work.

Relative oxygen uptake is expressed in $mL{\cdot}kg^{-1}{\cdot}min^{-1}$. It reflects the ability to move one's body and is inversely related to body fatness (Mayo, Grantham, & Balasekaran, 2002; Balasekaran, 2003; Mayo, Grantham, & Balasekaran, 2003; Balasekaran et al., 2005; Lee & Balasekaran, 2010; Balasekaran, Govindaswamy, Chia, & Lim, 2011; Gupta, Balasekaran, Govindaswamy, Chia, & Lim, 2011; Balasekaran, Gupta, Govindaswamy, Wang, & Bakri, 2014; Balasekaran, Boey, Hui, Govindaswamy, Ng, & Lim 2018; Balasekaran, Mayo, & Lim, 2019).

$$\textbf{Relative } VO_2\ (mL{\cdot}kg^{-1}{\cdot}min^{-1}) = (\textbf{Absolute } VO_2\ (L{\cdot}min^{-1}) \times 1000) \div \textbf{body weight}$$

Oxygen consumption can be expressed in metabolic equivalent (MET). MET can be used to determine an individual's functional capacity and help in prescribing exercise.

$$1\ MET = 3.5\ mL{\cdot}kg^{-1}{\cdot}min^{-1}$$

1 MET/3.5 $mL{\cdot}kg^{-1}{\cdot}min^{-1}$ is a typical resting value of oxygen consumption for humans.

1. **Participant A (100 kg body weight) had a VO_{2max} of 2.1 litres and Participant B (50 kg body weight) had a VO_{2max} of 2.1 litres. Explain the data and determine who is fit when you compare both participants.**

2. **What is 40 $mL{\cdot}kg^{-1}{\cdot}min^{-1}$ in METs?**

The Oxygen Cost of Breathing

Respiratory quotient (RQ) refers to the ratio of the amount of carbon dioxide produced divided by the amount of oxygen consumed at the cellular level.

Respiratory exchange ratio (RER) (Table 2) refers to the ratio of the volume of carbon dioxide produced divided by the volume of oxygen consumed at total body level. Although RER and RQ definitions differ, they are used interchangeably, as found in the literature.

$$\textbf{RQ or RER} = \textbf{VCO}_2/\textbf{VO}_2$$

(VCO_2: volume of carbon dioxide; VO_2: volume of oxygen)

Criteria for Determining a Maximal Test During a Treadmill Test (American College of Sports Medicine Guidelines (ACSM, 2018))

1. Levelling of VO_{2max}
   - VO_{2max} should level off near the end of the test and show no further increase or only slight increase with higher workloads.
2. Time
   - A good maximal test usually ends in 12–13 minutes.

Table 2. Caloric equivalence of the RER and % kcal from carbohydrates and fats (adapted from Wilmore & Costill, 1999).

	Energy	% kcal	
RER	**kcal/LO_2**	**Carbohydrates**	**Fats**
0.71	4.69	0.0	100.0
0.75	4.74	15.6	84.4
0.80	4.80	33.4	66.6
0.85	4.86	50.7	49.3
0.90	4.92	67.5	32.5
0.95	4.99	84.0	16.0
1.00	5.05	100.0	0.0

3. RER value >1.0.
   - RER = the respiratory exchange ratio measured at the mouth. It is the ratio of carbon dioxide (VCO_2) divided by the volume of oxygen (VO_2) consumed (Table 2).
   - An RER value > 1.0 indicates that the lungs are hyperventilating in order to buffer the lactic acid in the blood. This indicates that a person is probably working hard enough to attain maximal test. An increase in blood lactic acid represents an imbalance in production and metabolism, which is brought on during intense exercise. The body is trying to compensate for the lack of energy by breaking down glycogen (anaerobic glycolysis). Lactic acid is the end of this pathway (Chapter 3).
4. Heart Rate (HR)
   - For a maximal test, HR that should be at or above the age predicted maximum HR (220 – age).
   - HR rises in a linear fashion as oxygen uptake increases and attains a plateau just prior to the achievement of VO_{2max} during a graded exercise test.
5. Blood lactate > 8 $mmol{\cdot}L^{-1}$
   - Blood lactate should be above 8 $mmol{\cdot}L^{-1}$.
6. Rate of Perceived Exertion (RPE)
   - Robertson's OMNI RPE scale (> 9) (Robertson, 2004).
   - Borg scale (> 17 or > 9, depending on which Borg scale used).
7. VO_2 Plateau
   - A change in VO_2 less than 2.1 $mL{\cdot}kg^{-1}{\cdot}min^{-1}$ or 150 $mL{\cdot}min^{-1}$ from one workload to the next.

VO_2 Issues

The reason why a person attains true maximal test may not be due to oxygen limitation but instead due to several other factors (Plowman & Smith, 2003) such as:

- Recruitment of fast-twitch fibres
- Blood pH levels during maximal test, which may influence energy transport (blood pH might drop due to lactic acid) (Chapter 3 Figure 4)
- Ventilation responses (breathing rate increases and is unable to match oxygen supply needed for the muscle. There is a need to expire carbon dioxide to free bicarbonates needed for lactate buffering.) (Chapter 3 Figures 3 & 4)
- Muscle cross-bridge activation, which is hampered by lactic acid
- Economy of movement (other muscles are recruited to aid in performance due to fatigue of major muscle groups)

The above factors can explain fatigue at the end of a maximally exerted run.

The person with the highest VO_2 does not always win or perform the best. Why?

a) Genetics or natural endowment (fibre types — refer to Chapter 6).
b) The ability to work at a higher percentage of their VO_{2max} due to training and the ability to tolerate pain/discomfort associated with it.
c) Training can increase VO_{2max} by approximately 25% and raise mitochondrial content by about 100%. These are central and peripheral training adaptations and will be explained in Chapter 7.

1. **The following is an example of an individual's aerobic metabolic response during an incremental treadmill test (Table 3).**

Gender: Male	**Ambient Temperature: 23°C**
Age: 22 yr	**Barometric Pressure: 739 mmHg**
Weight: 66 kg	**Relative Humidity: 13%**

(a) The participant above did a submaximal to exhaustion protocol test. Did he attain the VO_{2max} criteria as outlined in this chapter? Explain and describe the energy sources he used during the exercise test.

(b) What was his VO_{2max}? Provide comments on his VO_{2max} value.

2. **Why was the maximal test conducted for 28 mins? Is this possible to determine VO_{2max}?**

3. **Table 3 also indicated that the participant started the maximal oxygen test during stage 1 at 0.99 RER. Is this RER possible? How did he manage to finish with a high oxygen consumption of 77.52 $mL{\cdot}kg^{-1}{\cdot}min^{-1}$ when he started at a high RER?**

Table 3. Aerobic metabolic responses during an incremental treadmill test (Modified Blake Protocol) (adapted from Plowman & Smith, 2003).

Time (min)	V_ESTPD ($L \cdot min^{-1}$)	O_2%	CO_2%	VO_2 ($L \cdot min^{-1}$)	VCO_2 ($L \cdot min^{-1}$)	VO_2 ($mL \cdot kg^{-1} \cdot min^{-1}$)	RER	HR ($beats \cdot min^{-1}$)
2	25.21	16.47	4.45	1.12	1.10	16.96	0.99	82
3	27.31	16.36	4.54	1.25	1.22	18.93	0.98	84
4	29.06	16.30	4.62	1.33	1.30	20.30	0.98	100
5	29.75	16.18	4.71	1.41	1.39	21.36	0.98	98
6	33.62	16.09	4.84	1.62	1.60	24.54	0.99	108
7	30.17	16.08	4.99	1.45	1.49	21.96	1.03	105
8	34.37	15.99	5.03	1.68	1.72	25.45	1.01	120
9	37.92	16.09	5.03	1.81	1.89	27.42	1.04	117
10	37.17	16.07	4.96	1.79	1.83	27.12	1.02	115
11	38.84	15.89	5.03	1.68	1.72	25.45	1.01	120
12	39.56	15.82	5.10	2.01	2.00	30.60	0.99	122
13	39.84	15.57	5.23	2.15	2.07	32.57	0.96	124
14	43.37	15.51	5.30	2.36	2.28	35.75	0.96	132
15	46.64	15.62	5.41	2.45	2.50	37.27	1.01	144
16	47.37	15.68	5.42	2.45	2.54	37.27	1.03	138
17	50.87	15.85	5.21	2.55	2.62	38.78	1.02	143
18	51.53	15.52	5.48	2.78	2.80	42.12	1.01	146
19	55.12	15.73	5.38	2.83	2.95	42.87	1.03	155
20	56.84	15.74	5.32	2.91	3.00	44.24	1.02	158
21	58.54	15.63	5.38	3.08	3.12	46.81	1.01	162
22	59.95	15.58	5.43	3.19	3.23	48.33	1.01	167
23	68.06	15.59	5.47	3.61	3.70	54.69	1.02	180
24	78.06	15.69	5.55	4.01	4.30	60.90	1.07	181
25	85.18	15.70	5.62	4.36	4.75	66.21	1.08	188
26	94.24	15.78	5.75	4.68	5.39	71.06	1.14	192
27	112.96	16.29	5.51	4.98	6.18	75.45	1.24	196
28	139.04	16.97	4.94	5.11	6.83	77.52	1.34	200

1. **Participant A has a VO_{2max} of 80 mL·kg^{-1}·min^{-1} and can run at 80% of his VO_{2max} with considerable ease. Explain why.**

2. **Participant A has a VO_{2max} of 80 mL·kg^{-1}·min^{-1} and Participant B has a VO_{2max} of 70 mL·kg^{-1}·min^{-1}. Participant A runs at 80% of his VO_{2max} and Participant B runs at 60% of his VO_{2max}. Explain the differences you will expect for the individuals in terms of running speed and sustenance of duration in running a long-distance endurance event.**

3. **Explain this statement: Increase in running speed can occur without increasing VO_{2max} values.**

Two participants have a VO_{2max} of 70 mL·kg^{-1}·min^{-1} and 80 mL·kg^{-1}·min^{-1}, respectively.

It would be difficult to predict who would win a 10-km race or do well in a 2.4-km race based on this information. The person with a VO_{2max} of 70 mL·kg^{-1}·min^{-1} with a better running economy could win the race.

Running economy refers to a steady state of VO_2 for a particular running velocity, where the body is supposed to be more efficient and does not take up as much oxygen for the same amount of work (see Figure 4). It may be an important characteristic in determining success between homogenous individuals, and hence it is essential to train one's running economy from young, regardless of sport (Balasekaran, 1993; Balasekaran, 2001) (Refer to *Applied Physiology of Exercise Laboratory Manual* Laboratory Session 15, Balasekaran, Govindaswamy, Lim, Boey, & Ng, 2021).

Relationship between running economy and performance time

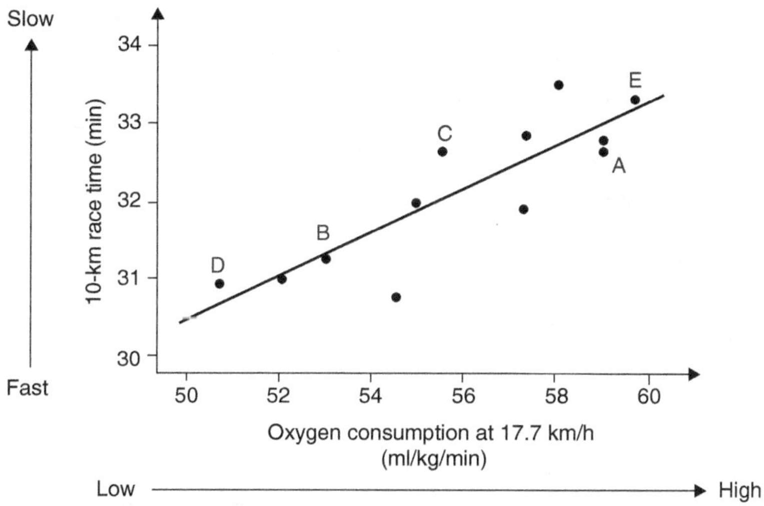

Figure 4. Relationship between 10-km race time (min) and oxygen consumption at 17.7 km•hr^{-1} (mL•kg^{-1}•min^{-1}) (adapted from Conley & Krahenbuhl, 1980).

1. Who has the best running economy (A, B, C, D, or E)? Why? (Figure 4)

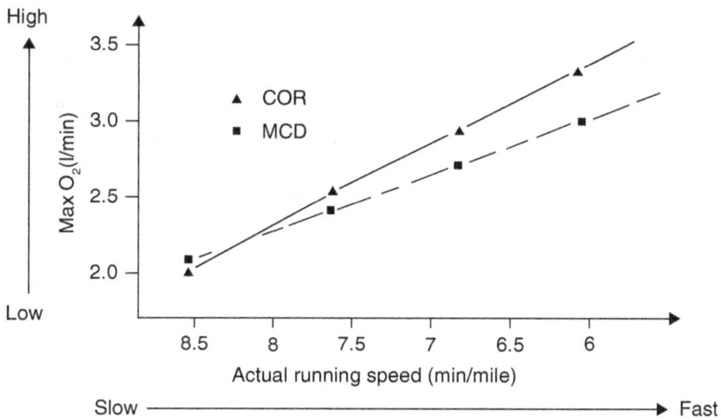

Figure 5. Oxygen requirements for Jim McDonagh (squares) and Ted Corbitt (triangles) while running at various speeds (adapted from Costill, 1986).

2. Who will have a better running performance in a long-distance endurance race? COR or MCD? Why? (Figure 5)

Aerobic Capacity and Running Performance

Table 4. Ratings of maximal oxygen uptake (mL•kg × min) for young men and women. The values in the right-hand column (potential 10-kilometre time) offer an estimate of the runner's running potential (min:sec) (adapted from Costill, 1986).

	Aerobic Capacity	Potential 10-km Time (min:sec)	
High ↑	Above 70 ml/kg × min	33:00 or faster	↑ Fast
	65 to 69 ml/kg × min	36:15 to 33:40	
	60 to 64 ml/kg × min	39:30 to 36:50	
	55 to 59 ml/kg × min	42:45 to 40:10	
	50 to 54 ml/kg × min	46:00 to 43:25	
	45 to 49 ml/kg × min	49:15 to 46:40	
	40 to 44 ml/kg × min	52:30 to 49:50	
Low	Below 39 ml/kg × min	53:10 or slower	Slow

1. **Can Table 4 be used for children, adolescents, and teenagers? If yes, please explain how to use it for each category.**

2. **Can you use Table 4 to train an elite soccer team? If yes, explain how you will utilise it for your elite soccer team. Specify the gender and explain how you would utilise the physiological differences in gender (refer to page 53) and fitness (Hint: women have lower aerobic capacity compared to men) to train an elite soccer team.**

VO_{2max} for Other Sports

Table 5. Typical values for maximal oxygen consumption in highly-trained young adult athletes (adapted from Wilmore & Costill, 2005).

Running Sports	**Treadmill ($mL{\cdot}kg^{-1}{\cdot}min^{-1}$)**	**Mean**	**Usual Range**
Distance Runners	Male	75	70–80
	Female	65	60–75
Soccer Players	Male	60	55–65
	Female	50	45–55
Field Hockey Players	Male	60	55–65
	Female	50	45–55
Basketball Players	Male	55	50–60
	Female	45	40–55
Australian Football Players	Male	60	55–65
Rowing	**Rowing Ergometer ($L{\cdot}min^{-1}$)**		
	Heavyweight Male	5.5	5.0–6.2
	Heavyweight Female	3.8	3.3–4.4
Cycling	**Cycling Ergometer ($L{\cdot}min^{-1}$)**		
Track	Male	5.3	4.5–6.0
	Female	3.3	2.6–4.0
Road	Male	5.4	4.5–6.1
	Female	3.5	3.0–4.0
Road ($mL{\cdot}kg^{-1}{\cdot}min^{-1}$)	Male	75	6.5–8.0
	Female	65	5.5–7.0

Table 6. Standards for 15-minute run for team game players (m) (adapted from Caplan, 2007).

	Male	**Female**
Very Good	>4200	>3800
Good	3900–4200	3500–3800
Average	3600–3899	3250–3499
Fair	3300–3599	3000–3249
Poor	<3300	<3000

Table 7 is an anaerobic test, which is short in duration and high in intensity. This is opposite of an aerobic test like a 1.6-km run (see equation on page 49), 2.4-km (1.5-mile) run (Table 8), and 15-minute run (Table 6 and equation on page 48), which are longer in duration and lower in intensity.

Table 7. Anaerobic endurance test for 400-metre run(s) (adapted from Caplan, 2007).

	Male	Female
Very Good	<53.0	<59.0
Good	53.0–55.9	59.0–61.9
Average	56.0–58.9	62.0–64.9
Fair	59.0–62.0	65.0–68.0
Poor	>62.0	>68.0

1. **The timings in Table 7 are suitable for adults. How can you use it for teenagers (17 to 19 years old) to test their anaerobic endurance?**

2. **For adolescents below the age of 17, how will you use the timings in Table 7 to test their anaerobic endurance?**

Estimation of VO_{2max}

In addition to Table 6 to assess fitness, you can estimate VO_{2max} from a 15-min run and 1.6-km run as indicated below. You may refer to Table 5 to determine your fitness level according to various sports. You can also refer to Tables 13, 14, 15, and 16.

To estimate from a 15-min run:

Example 1. A soccer player who runs 3,800 metres in 15 minutes:

$VO_{2max} = 33.3 + (\text{distance covered}/15 - 133) \times 0.172$

$VO_{2max} = 33.3 + (3800/15 - 133) \times 0.172$

The estimated VO_{2max} of a soccer player who runs 3,800 metres in 15 minutes is **53.9 mL·kg⁻¹·min⁻¹**.

To estimate from a 1.6-km run:

Example 2. A distance runner who covers 1.6 km in 5 minutes and 6 seconds:

$VO_{2max} = 133.61 - (13.89 \times \text{time for run})$

$VO_{2max} = 133.61 - (13.89 \times 5.1)$

The estimated VO_{2max} of a distance runner who covers 1.6 km in 5 minutes and 6 seconds is **62.8 mL·kg⁻¹·min⁻¹**.

Example: A 20-year-old female runs the 1.5-mile (2.4-km) course in 12 minutes and 40 seconds. Table 8 shows a VO_{2max} of 39.8 $mL \cdot kg^{-1} \cdot min^{-1}$ for a time of 12:40. This VO_{2max} places her in the excellent cardiorespiratory fitness category (Tables 14 & 16).

Table 8. Assessing maximal oxygen uptake for a 1.5-mile run test (adapted from Cooper, 1968; Pollock, Wilmore, & Fox, 1978; Wilmore & Costill, 1988).

Time	VO_{2max} $mL \cdot kg^{-1} \cdot min^{-1}$	Time	VO_{2max} $mL \cdot kg^{-1} \cdot min^{-1}$
6:10	80.0	12:50	39.2
6:20	79.0	13:00	38.6
6:30	77.9	13:10	38.1
6:40	76.7	13:20	37.8
7:00	74.0	13:30	37.2
7:10	72.6	13:40	36.8
7:20	71.3	13:50	36.3
7:30	69.9	14:00	35.9
7:40	68.3	14:10	35.5
7:50	66.8	14:20	35.1
8:00	65.2	14:30	34.7
8:10	63.9	14:40	34.3
8:20	62.5	14:50	34.0
8:30	61.2	15:00	33.6
8:40	60.2	15:10	33.1
8:50	59.1	15:20	32.7
9:00	58.1	15:30	32.2
9:10	56.9	15:40	31.8
9:20	55.9	15:50	31.4
9:30	54.7	16:00	30.9
9:40	53.5	16:10	30.5
9:50	52.3	16:20	30.2
10:00	51.1	16:30	29.8
10:10	50.4	16:40	29.5
10:20	49.5	16:50	29.1

(Continued)

Table 8. *(Continued)*

Time	VO_{2max} $mL \cdot kg^{-1} \cdot min^{-1}$	Time	VO_{2max} $mL \cdot kg^{-1} \cdot min^{-1}$
10:30	48.6	17:00	28.9
10:40	48.0	17:10	28.5
10:50	47.4	17:20	28.3
11:00	46.6	17:30	28.0
11:10	45.8	17:40	27.7
11:20	45.1	17:50	27.4
11:30	44.4	18:00	27.1
11:40	43.7	18:10	26.8
11:50	43.2	18:20	26.6
12:00	42.3	18:30	26.3
12:10	41.7	18:40	26.0
12:20	41.0	18:50	25.7
12:30	40.4	19:00	25.4
12:40	39.8		

You can refer to Table 5 for categorising of various sports.

Table 9 is an anaerobic test, which is short in duration and high in intensity. This is the opposite of an aerobic test like a 1.6-km run (see equation on page 49), 2.4-km (1.5-mile) run (Table 8), and 15-minute run (Table 6 and equation on page 48), which are longer in duration and lower in intensity.

Table 9. Anaerobic power and speed tests results (adapted from Caplan, 2007).

Tests	Male	Female
Vertical Jump (cm)		
Very Good	>67	>57
Good	60–67	50–56
Average	52–59	43–49
Fair	44–51	36–42
Poor	<44	<36
35-metre Dash (s)		
Very Good	<4.80	<5.30
Good	4.80–5.09	5.30–5.59
Average	5.10–5.29	5.60–5.89
Fair	5.30–5.60	5.90–6.20

(Continued)

Table 9. *(Continued)*

Tests	Male	Female
5-0-5 Test* (s)		
Very Good	<2.20	<2.40
Good	2.20–2.39	2.40–2.54
Average	2.30–2.49	2.55–2.74
Fair	2.50–2.60	2.75–2.90
Poor	>2.60	>2.90

**This test involves running 5 metres to and from an end line, from a moving start (use of timing gates will be more accurate than a stopwatch).*

Tables 10 and 11 are flexibility and body composition tests, respectively (Balasekaran et al., 2016). These are tests which can be used in conjunction with an aerobic test. A person with higher VO_{2max} will have better flexibility and lower body composition.

Flexibility and Body Composition Tests

Table 10. Sit-and-reach test (cm) (adapted from Caplan, 2007).

	Male	Female
Very Good	>13.0	>15.0
Good	10.1–13.0	12.1–15.0
Average	6.1–10.0	8.1–12.0
Fair	1.0–6.0	3.0–8.0
Poor	<1.0	<3.0

Table 11. Skinfold thickness (mm) (adapted from Telford et al., 1988).

		Mean	Range
Basketball	Male*	76	40–89
	Female**	84	54–133
Cycling (Track)	Male	48	33–79
	Female	75	56–104
Gymnastics	Male	39	29–57
	Female	45	34–55
Field Hockey	Male	49	34–68
	Female	71	55–92
Rowing	Male	55	36–76
	Female	94	55–133
Swimming	Male	50	38–75
	Female	66	47–90
Track and Field (Sprinting)	Male	42	32–69
	Female	67	39–97
Middle-distance Running	Male	38	32–44
	Female	63	31–90
Throwing	Male	107	45–224
	Female	135	103–220

*Males: sum of 8 skinfolds **Females: sum of 7 skinfolds

(Refer to *Applied Physiology of Exercise Laboratory Manual* Laboratory Session 1, Balasekaran, Govindaswamy, Lim, Boey, & Ng, 2021)

Velocity at VO_{2max}

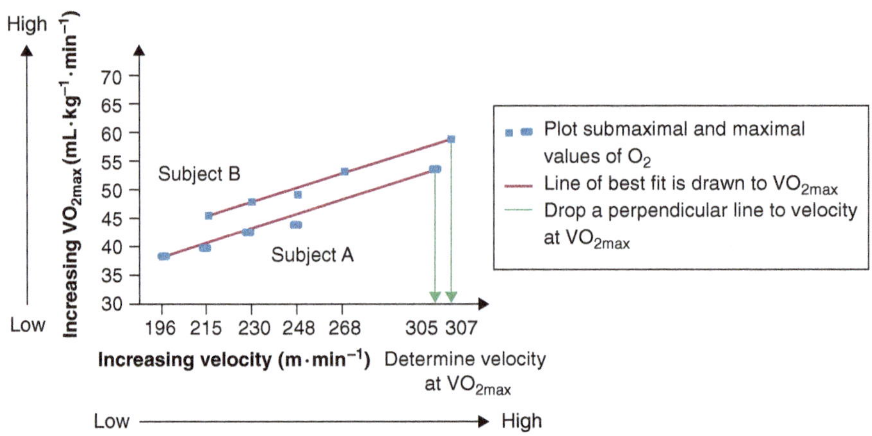

Figure 6. Predicted velocity at VO_{2max} (adapted from Plowman & Smith, 2003).

1. **Can velocity at VO_{2max} be used for training? If yes, explain. (Figure 6) (Refer to *Applied Physiology of Exercise Laboratory Manual* Laboratory Session 12, Balasekaran, Govindaswamy, Lim, Boey, & Ng, 2021).**

2. **Is velocity at VO_{2max} the same as maximal aerobic speed? (Figure 6) (Refer to *Applied Physiology of Exercise Laboratory Manual* Laboratory Session 12, Balasekaran, Govindaswamy, Lim, Boey, & Ng, 2021).**

Other Factors that May Influence VO_{2max}

1. Age
2. Body size and muscle mass
3. Whole body, arms, legs
4. Gender: females have approximately 15–25% lower aerobic capacity than males due to:
    - Relatively higher body fat
    - Relatively lower lean body weight
    - Lesser blood volume
    - Smaller heart size
    - Lesser haemoglobin (Hb)
5. Specificity of testing
6. Detraining results in a rapid loss in central relative to peripheral functions. This statement is debatable and depends on the history of the athlete.
    - If the athlete has trained/exercised for a long time, the chances of losing the central relative to the peripheral might be lesser.
    - For an unfit individual, the effects of detraining might be more pronounced in the peripheral compared to the central.
7. Level of activity: importance of regular physical activity

8. Smoking: affects central, peripheral, and transport
   - Increases airway resistance ⇒ increased oxygen cost of breathing
   - Oxygen, carbon monoxide, and Hb levels drop as the ability of Hb to transport oxygen to working muscle decreases
   - Carbon monoxide has a higher affinity to Hb than oxygen
   - Arteries become less elastic
9. Functional limitations in disease states — lower VO_{2max}
   - Myocardial ischemia: myocardial oxygen demands exceed oxygen consumption
   - Bed rest
   - Other pathophysiological considerations: pulmonary dysfunction

Healthy, young, and relatively untrained adults can increase their VO_{2max} by 15–20% or more, depending on their initial level of fitness. The higher the fitness level, the harder it is to increase VO_{2max}. If the initial fitness level is low, the magnitude of increase is high (Balasekaran et al., 2005; Gupta & Balasekaran, 2013; Ali, Balasekaran, Hoon, & Gerald, 2017).

Periodic and regular testing of aerobic power can help determine:

1. The suitability of an athlete for a given type of sport or a specific role in a sport
2. The kind of emphasis that should be placed during aerobic training
3. The type of aerobic training which should be employed
4. The effect of a given fitness conditioning programme on maximal aerobic power
5. The rate at which an athlete is improving, or the rate at which a fitness conditioning programme is eliciting change
6. The running pace at which an athlete should compete
7. Whether an athlete is suffering some decline in capacity due to biological age, nutritional, or medical factors
8. To assess peripheral or central physiological adaptation deficiency in an athlete (refer to Chapter 7)

Relevance of Maximal Aerobic Power

VO_{2max} is important for sports or events with a duration longer than 2 minutes. However, aerobic fitness is needed for all types of sports.

Racquet sports and most team sports involve bursts of highly intense energy release, separated by periods of lower intensity. Its recovery portion is an oxidative process.

Athletes in sports demanding sustained effort in excess of 2 minutes have higher aerobic power compared to those in shorter or intermittent type of sports (Balasekaran, 1999; Grantham & Balasekaran, 2004).

Highly Relevant	**Relevant**	**Of Little Relevance**
Track (400 m upwards)	Judo	Jumping & Throwing
Orienteering	Team sports	Diving
Swimming (100 m upwards)	Gymnastics	Weightlifting
Rowing	Canoeing	Shooting
Canoeing	Racquet sports	Archery
Cycling	Taekwondo	Sailing
Boxing	Tai Chi	High jump
Wrestling	Wushu	Discus
Mid-field soccer player		Javelin
Marathon		Long jump

Sports listed under 'Of Little Relevance' will still require a certain level of aerobic fitness to maintain good cardiovascular health.

Total Body Oxygen Kinetic Chain

1. **What is a limiting factor in performance? Oxygen take up/uptake, transport, or utilisation? Explain.**

2. **Do we need to train all 3 components to be physically fit for competition?**

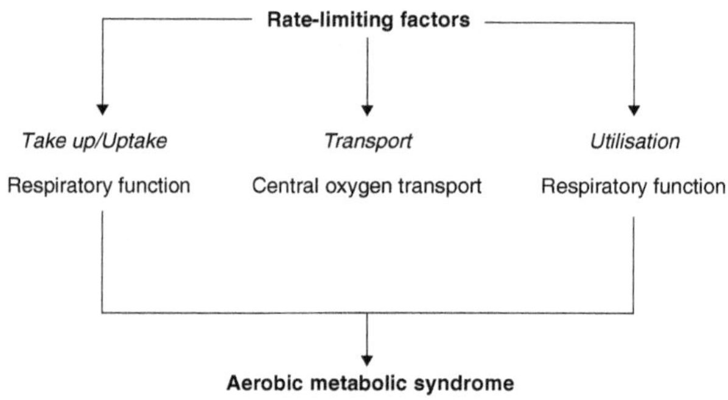

3. **Which causes fatigue — the inability to take up or the inability to utilise oxygen? Explain.**

4. **When you run a long endurance run, can you feel which is causing your fatigue — the inability to take up or the inability to utilise oxygen? Explain.**

5. **A person completes a VO_{2max} test on a treadmill and attains all criteria for a maximal test but does not reach a maximum HR of (220 – age). Can this happen? Explain why this person has such a result.**

Hint: Increase in maximal work rate and capacity for the endurance exercise result from a unique combination of adaptations in the cardiovascular system (central) and skeletal muscles (peripheral) (Figure 7). This will be further elaborated in Chapter 7.

Another Diagram Showing Areas of Possible Limitations to Maximal Oxygen Consumption

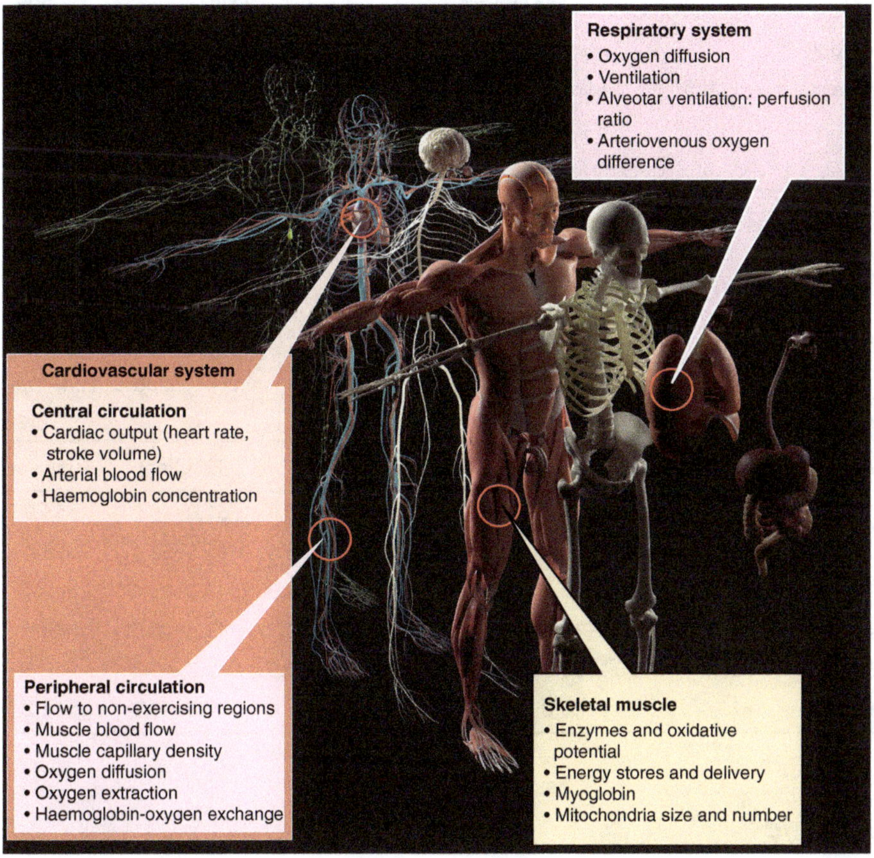

Figure 7. Possible limitations to VO_{2max} (adapted from Plowman & Smith, 1997).

Biomotor Abilities of a Sport

Table 12 dissects a sport like Australian football and allows the coach to conduct training according to the specific demands of the sport.

Table 12. Estimates of fitness and skill demands of Australian football (adapted from Woodman & Pyke, 1991).

Energy Systems	20%	Muscular Fitness	20%	Skills	60%
Aerobic	50%	General strength	15%	Individual skills	20%
Anaerobic	50%	General muscular endurance	25%	Small-group skills	30%
		Specific power	50%	Team plays	50%
		Flexibility	10%		

VO_{2max} Age-Based Criteria

Males

Table 13. VO_{2max} values for males based on age ($mL \cdot kg^{-1} \cdot min^{-1}$) (adapted from The Physical Fitness Specialist Certification Manual).

Age	Very Poor	Poor	Fair	Good	Excellent	Superior
13–19	<35.0	35.0–38.3	38.4–45.1	45.2–50.9	51.0–55.9	>55.9
20–29	<33.0	33.0–36.4	36.5–42.4	42.5–46.4	46.5–52.4	>52.4
30–39	<31.5	31.5–35.4	35.5–40.9	41.0–44.9	45.0–49.4	>49.4
40–49	<30.2	30.2–33.5	33.6–38.9	39.0–43.7	43.8–48.0	>48.0
50–59	<26.1	26.1–30.9	31.0–35.7	35.8–40.9	41.0–45.3	>45.3
60+	<20.5	20.5–26.0	26.1–32.2	32.3–36.4	36.5–44.2	>44.2

Females

Table 14. VO_{2max} values for females based on age ($mL \cdot kg^{-1} \cdot min^{-1}$) (adapted from The Physical Fitness Specialist Certification Manual).

Age	Very Poor	Poor	Fair	Good	Excellent	Superior
13–19	<25.0	25.0–30.9	31.0–34.9	35.0–38.9	39.0–41.9	>41.9
20–29	<23.6	23.6–28.9	29.0–32.9	33.0–36.9	37.0–41.0	>41.0
30–39	<22.8	22.8–26.9	27.0–31.4	31.5–35.6	35.7–40.0	>40.0
40–49	<21.0	21.0–24.4	24.5–28.9	29.0–32.8	32.9–36.9	>36.9
50–59	<20.2	20.2–22.7	22.8–26.9	27.0–31.4	31.5–35.7	>35.7
60+	<17.5	17.5–20.1	20.2–24.4	24.5–30.2	30.3–31.4	>31.4

Aerobic Power Tests

Males

Table 15. Aerobic power tests for men (adapted from American College of Sports Medicine, 2018).

	Age (20–29) N = 1675				Age (30–39) N = 7094			
% Mile	Bike Treadmill (time)	VO_{2max} (mL•kg^{-1}• min^{-1})	12-min Run Distance (miles)	1.5-mile Run (time)	Bike Treadmill (time)	VO_{2max} (mL•kg^{-1}• min^{-1})	12-min Run Distance (miles)	1.5-mile Run (time)
99	≥30:20	≥58.79	≥1.94	≤7:29	≥29:00	≥58.86	≥1.89	≤7:11
95	27:00	53.97	1.81	8:13	26:00	52.53	1.77	8:44 Superior
90	25:11	51.35	1.74	9:09	24:30	50.36	1.71	9:30
85	24:00	49.64	1.69	9:45	23:00	48.20	1.65	10:16
80	23:00	48.20	1.65	10:16	22:00	46.75	1.61	10:47 Excellent
75	22:10	46.99	1.62	10:42	21:00	45.31	1.57	11:18
70	22:00	46.75	1.61	10:47	20:30	44.59	1.55	11:34
65	21:00	45.31	1.57	11:18	20:00	43.87	1.53	11:49
60	20:15	44.23	1.54	11:41	19:00	42.42	1.49	12:20 Good
55	20:00	43.87	1.53	11:49	18:25	41.58	1.47	12:38
50	19:03	42.49	1.50	12:18	18:00	40.98	1.45	12:51
45	19:00	42.42	1.49	12:20	17:00	39.53	1.41	13:22
40	18:00	40.98	1.45	12:51	16:32	38.86	1.39	13:36 Fair
35	17:30	40.26	1.43	13:06	16:00	38.09	1.37	13:53
30	17:00	39.53	1.41	13:22	15:30	37.37	1.35	14:08
25	16:00	38.09	1.37	13:53	15:00	36.65	1.33	14:24
20	15:20	37.13	1.34	14:13	14:06	35.35	1.29	14:52 Poor
15	15:00	36.65	1.33	14:24	13:10	34.00	1.25	15:20
10	13:30	34.48	1.27	15:10	12:09	32.53	1.21	15:52
5	11:30	31.57	1.19	16:12	11:00	30.87	1.17	16:27 Very Poor
1	≤8.23	≤27.09	≤1.06	≥17:48	≤8:00	≤26.54	≤1.13	≥18:00

Females

Table 16. Aerobic power tests for women (adapted from American College of Sports Medicine, 2018).

	Age (20–29) N = 764				Age (30–39) N = 2049			
% Mile	Bike Treadmill (time)	VO_{2max} (mL•kg^{-1}• min^{-1})	12-min Run Distance (miles)	1.5-mile Run (time)	Bike Treadmill (time)	VO_{2max} (mL•kg^{-1}• min^{-1})	12-min Run Distance (miles)	1.5-mile Run (time)
99	≥26:21	≥53.03	≥1.78	≤8:33	≥23:22	≥48.73	≥1.66	≤10:05
95	22:00	46.75	1.61	10:47	20:00	43.87	1.53	11:49 Superior
90	20:12	44.15	1.54	11:43	18:00	40.98	1.45	12:51
85	19:00	42.42	1.49	12:20	17:30	40.26	1.43	13:06
80	18:00	40.98	1.45	12:51	16:20	38.57	1.38	13:46 Excellent
75	17:00	39.53	1.41	13:22	15:30	37.37	1.35	14:08
70	16:00	38.09	1.37	13:53	15:00	36.65	1.33	14:24
65	15:30	37.37	1.35	14:08	14:10	35.44	1.29	14:50
60	15:00	36.65	1.33	14:24	13:35	34.60	1.27	15:08 Good
55	14:39	36.14	1.31	14:35	13:10	33.85	1.26	15:20
50	14:00	35.20	1.29	14:55	13:00	33.76	1.25	15:26
45	13:30	34.48	1.27	15:10	12:10	32.41	1.22	15:47
40	13:00	33.76	1.25	15:26	12:00	32.31	1.21	15:57 Fair
35	12:17	32.72	1.22	15:48	11:09	31.09	1.17	16:23
30	12:00	32.31	1.21	15:57	10:45	30.51	1.16	16:35
25	11:03	30.94	1.17	16:26	10:00	29.93	1.13	16:58
20	10:50	30.63	1.16	16:33	9:30	28.70	1.11	17:14 Poor
15	10:00	29.43	1.13	16:58	9:00	27.98	1.09	18:00
10	9:17	28.39	1.10	17:21	8:00	26.54	1.05	18:00
5	7:33	25.89	1.03	18:14	7:00	25.09	1.01	18:31 Very Poor
1	≤5.15	≤22.57	≤9.4	≥19:25	≤5:12	≤22.49	≤9.3	≥19:27

PACER Test/Multistage Fitness Test (Appendix A–C)

- Predicted VO_{2max}
- Good indication of aerobic fitness
- Easy to conduct with masses (appropriate in school settings)
- Small space (15 m for children, 20 m for adults) and little equipment (cones) required
- Easy-to-follow recorded audio and instructions
- Use as a fitness and conditioning workout
- Short maximal test (12 minutes)
- Use the level as performance, which is similar to a true VO_{2max} test using a treadmill

Limitations to PACER Test/Multistage Fitness Test

- Does not account for individual body weight.
- Researched with young college population — not accurate with other population (older/younger/different racial group) (there may be updated versions available).
- Sport-specific test: the test is beneficial for territorial, invasion, or net barrier games like soccer, basketball, netball, etc. as these games consist of stoppages/disruption while running (these players are used to multi-directional running). However, it might not be as beneficial for runners as they run continuously for long distances and are not use to stoppages. Long distance runners may also need longer distance/time to pick up speed as they are used to unidirectional running whereas the PACER test repeatedly stops their running abruptly (braking forces) and forces them to pick up speed immediately after a loss of speed.
- Obese children or individuals with orthopaedic problems are advised not to use this test. The excessive weight that they carry may affect their knees and joints as they need to repeatedly stop abruptly (braking forces) and continue running immediately (impact forces). It is more suitable for these individuals to use the one-mile walk instead (Appendix D).

1. **What is a good test to estimate VO_{2max} for a large class of 40 students? Explain why.**

2. **Can VO_{2max} increase always occur in a trained athlete? What are your deductions if there are no increases in VO_{2max} after a period of training?**

3. **Can a person with no/low increment in VO_{2max} from general preparation (conditioning) to a competition season still attain a fast timing in a 5,000-m or 10,000-m track race? If yes, what is the physiological rationale?**

4. **Is it a good idea to just rely on a maximal oxygen test to determine fitness or as an entry requirement to qualify for a soccer, rugby, or tennis team? Explain.**

Conclusion

Relevance of VO_{2max} testing: keeping all things in mind, VO_{2max} testing is still the best measure of functional oxidative capacity, which can be used in a variety of sports.

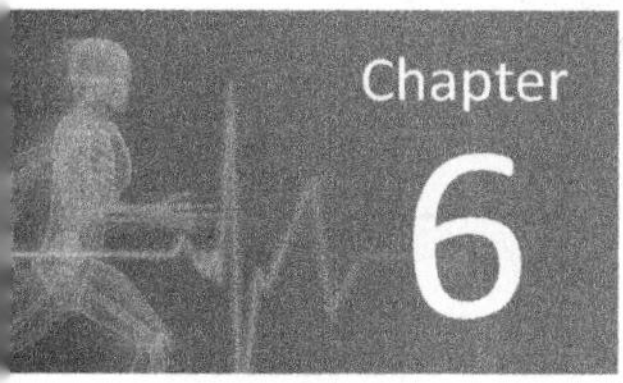

Fibre Types

Different muscles have different fibre type compositions; all muscles contain a mixture of fibre types; the proportions vary from muscle to muscle and also between individuals.

Table 1. Twitch properties — slow, fast (adapted from Plowman & Smith, 2003).

	Slow	Intermediate	Fast
Metabolic Properties	Oxidative	Oxidative and glycolytic	Glycolytic
Other Nomenclature	ST, Type 1	FTa, FTA, Type IIa	FTb, FTB, Type IIb, Type IIx
Motor Neuron Type	α_2	α_1	α_1
Motor Neuron Size	Small	Large	Large
Conduction Velocity	Slow	Fast	Fast
Recruitment Threshold	Low	High	High

Table 2. Structural and functional characteristics of muscle fibres (adapted from Plowman & Smith, 2003).

	Type I	Type II	
Contractile	ST	FTa	FTb
Metabolic	SO	FOG	FG
Structural Aspects			
Muscle fibre diameter	Small	Largest	Large
Mitochondrial density	High	High	Low

(*Continued*)

Table 2. *(Continued)*

	Type I	Type II	
Capillary density	High	Medium	Low
Myoglobin content	High	Medium	Low
Functional Aspects			
Twitch (contraction) time	Slow	Fast	Fast
Relaxation time	Slow	Fast	Fast
Force production	Low	High	High
Fatigability	Fatigue-resistant	Fatigable	Most fatigable
Metabolic Aspects			
Phosphocreatine stores	Low	High	High
Glycogen stores	Low	High	High
Triglyceride stores	High	Medium	Low
Myosin-ATPase activity	Low	High	High
Glycolytic enzyme activity	Low	High	High
Oxidative enzyme activity	Low	High	High

Approximate Values of Slow Twitch in Sedentary Males in Normal Population

Muscles	Slow Twitch %
Soleus	87
Tibialis Anterior	73
Biceps Femoris	67
Peroneus Longus	67
Gastrocnemius	55
Deltoid	52
Biceps Brachii	43
Vastus Lateralis	38
Triceps	32
Rectus Femoris	29
Orbicularis Oculi	15

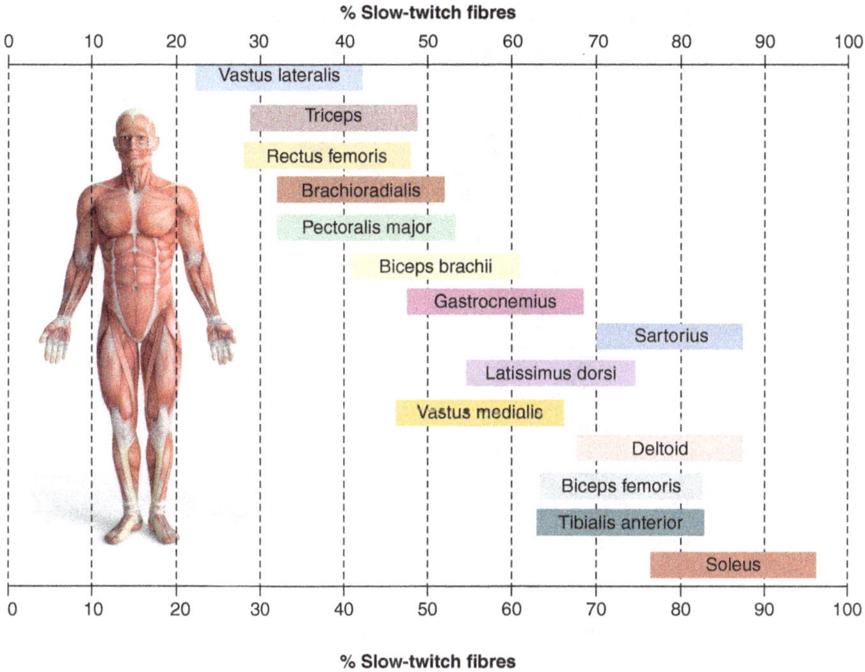

Figure 1. Percentage of slow-twitch fibres in sedentary males in normal population (adapted from Plowman & Smith, 2003).

> **1. What is the training effect of endurance, strength, and power on muscles fibres? (Hint: refer to Tables 1 and 2)**

Large Variability in Groups of People

Type I and Type II fibres are present in roughly equal numbers for most people (Figure 1). However, there are some people in which one type will predominate (Tables 1 & 2) (Riechman, Balasekaran, Roth, & Ferrell, 2004).

Fibre type difference helps to explain individual differences in performance and response to training (Table 3). It partly explains what training can and cannot do.

Table 3. Percentages and cross-sectional areas slow-twitch (ST) and fast-twitch (FT) fibres in selected muscles of male and female athletes (adapted from Wilmore & Costill, 1999).

					Cross-sectional area (μm^2)	
Athlete	**Gender**	**Muscle**	**%ST**	**%FT**	**ST**	**FT**
Sprint	M	Gastrocnemius	24	76	5,878	6,034
runners	F	Gastrocnemius	27	73	3,752	3,930
Distance	M	Gastrocnemius	79	21	8,342	6,485
runners	F	Gastrocnemius	69	31	4,441	4,128
Cyclist	M	Vastus lateralis	57	43	6,333	6,116
	F	Vastus lateralis	51	49	5,487	5,216
Swimmers	M	Posterior deltoid	67	33	—	—
Weightlifters	M	Gastrocnemius	44	56	5,060	8,910
	M	Deltoid	53	47	5,010	8,450
Triathletes	M	Posterior deltoid	60	40	—	—
	M	Vastus lateralis	63	37	—	—
	M	Gastrocnemius	59	41	—	—
Canoeists	M	Posterior deltoid	71	29	4,920	7,040
Shot-putters	M	Gastrocnemius	38	62	6,367	6,441

1. **If you have seen chicken meat, the thigh meat appears dark and the breast meat appears white. Name the muscle fibre type for the thigh and breast meat. How does the chicken use this muscle fibre type to evade predators? Explain.**

2. **How does the hyena attack its prey? Explain its attack tactics in relation to fibre types.**

3. **How does the cheetah hunt its prey? Explain its attack tactics in relation to fibre types.**

4. **By answering questions 1, 2, and 3, how does it relate to human beings' genetic potential in sports performance (nurture versus nature)? You can use the 100-m and marathon events as an example. (Hint: Figure 2)**

Percentage of Slow-Twitch Fibres in Athletes

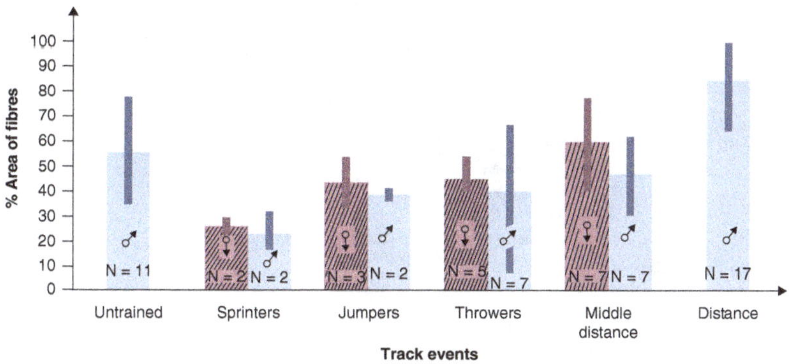

Figure 2. Percentage of gastrocnemius composed of slow-twitch fibres in male and female track athletes (adapted from Costill, 1986).

5. Which muscle fibre is activated first during a sprint start? Explain why. (Hint: Figures 3 & 4)

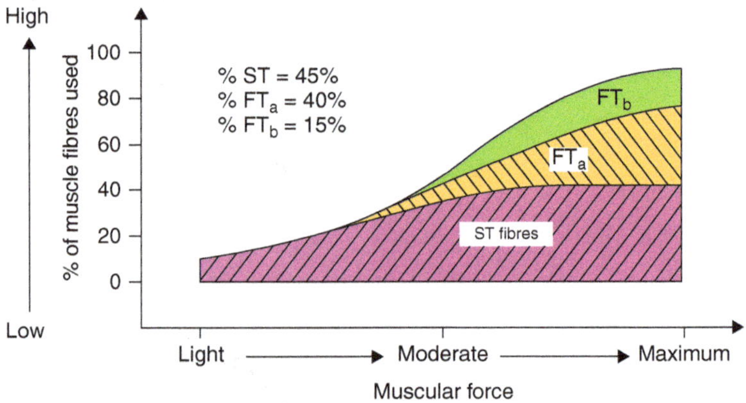

Figure 3. Ramp-like recruitment of muscle fibres with varying running speeds (adapted from Costill, 1986).

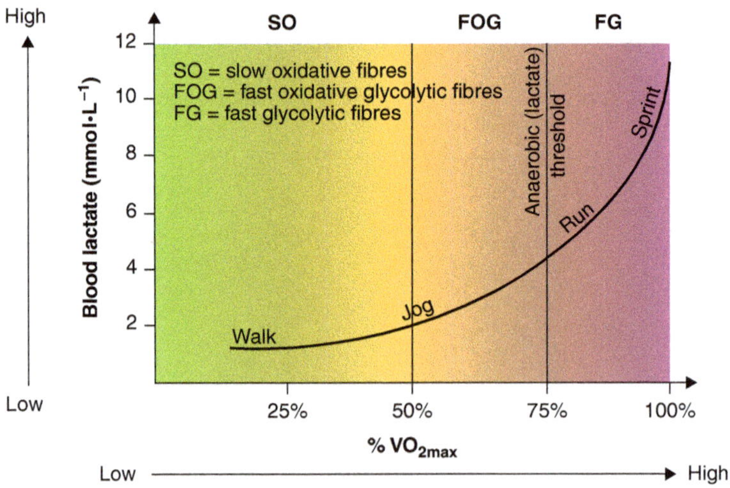

Figure 4. Anaerobic (lactate) threshold (adapted from Costil, 1986).

6. Which muscle fibre is activated for walking? Explain why. (Figures 3 & 4)

Fast oxidative glycolytic (FOG) fibres are recruited as exercise intensity increases. Further increases in exercise intensity result in the recruitment of fast glycolytic (FG) fibres. As FG fibres produces more lactate and fewer fibres are idle, they are unable to remove or take up lactate via the Cori cycle and eventually gluconeogenesis (Chapter 3 Figure 5), leading to lactate accumulation (Balasekaran, 2002) (Figure 4).

Exercise Intensity and Muscle Fibre Recruitment

Fibre Types and Performance

There are genetic or hereditary differences among athletes for fast-twitch versus slow-twitch muscle fibres:

- Some people have considerably more fast-twitch than slow-twitch fibres, and others have more slow-twitch fibres. This could, to some extent, determine the athletic capabilities of different individuals.

  E.g. Top-class sprinters tend to have a higher than normal proportion of Type II (75%–80%). Elite marathon runners tend to have up to 80% Type I.
- Training has not shown to change the relative proportions of fast-twitch and slow-twitch fibres however much an athlete wants to develop one type of athletic prowess over the other. Type IIa has shown to change its metabolic properties as a result of a specific type of training (endurance versus sprint). Slow-twitch fibres are also highly correlated with high VO_{2max} results.
- Determined almost entirely by genetic inheritance which in turn helps to determine which area of athletics is most suited to each person. Some people are born to be marathoners. Others are born sprinters and jumpers (Table 3 & Figure 2).

(adapted from Plowman & Smith, 2003; Powers & Howley, 2009)

1. **What are the estimate muscle fibre percentages for different sports? Choose 3 team sports that are not listed in Table 3 and explain. You can be specific with players' positions.**

2. **If you were a world-class sprinter, can you run a world-class marathon time, and vice versa? Explain.**

3. **Can we train and change the fibre type of a person? Explain.**

4. **Why do some students at secondary four (16 years old) become sprinters when they were long-distance runners in primary six (12 years old), or vice versa? The same question can apply to soccer positions where they were strikers when they were young and become midfielders when they become older. Did nature or nurture have an effect on their change? Explain.**

5. **Can fibre type differentiation be used as a tactic in a 5,000-m race or in a soccer team's selection of players for different positions to be used as a tactic to win? Explain.**

Chapter 7 Adaptations

Maximal Oxygen Uptake (VO_{2max})

Increased maximal oxygen uptake is the most significant physiological change that is induced by a well-designed endurance exercise-training programme.

- Allows a conditioned individual to perform at a higher workload for a longer period of time before being limited by fatigue, dyspnea, or chest pain (Figure 1)

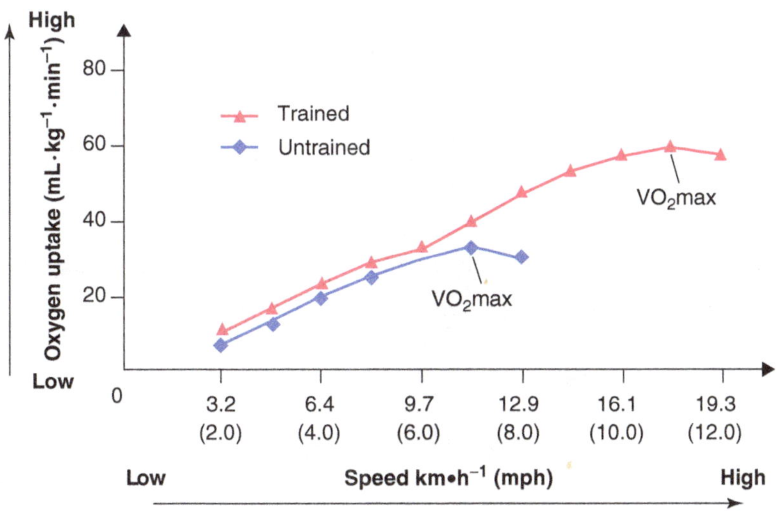

Figure 1. Relationship between exercise intensity and oxygen uptake in a trained and untrained man (adapted from Powers & Howley, 2009).

- Allows an individual to exercise at higher work rates in the trained state compared to the untrained state. A higher work rate is required by the untrained to attain the same relative rate of oxygen consumption (i.e. percentage of VO_{2max}) from the trained state (Figure 1).

The table below is an example of an oxygen consumption test which elicits a higher work rate by the untrained state in order to attain the same relative rate of oxygen consumption (i.e. percentage of VO_{2max}) as the trained state (Table 1).

Table 1. Example of pre- and post-oxygen consumption by an untrained individual before and after prolonged cardiovascular and physiological training.

	Pre	Post	Pre	Post
Speed	**VO_2**	**VO_2**	**$\%VO_2$**	**$\%VO_2$**
6.0	1.50	1.50	42.4	34.1
6.5	1.78	1.78	50.3	40.5
7.0	2.02	2.02	57.1	45.9
7.5	2.36	2.36	66.7	53.6
8.0	2.78	2.78	78.5	63.2
8.5	3.13	3.13	88.4	71.1
9.0	3.24	3.24	91.5	73.6
9.5	3.54	3.54	100	80.4
10.0		3.94		89.5
10.5		4.10		93.2
11.0		4.26		96.8
11.5		4.40		100

Increased maximal oxygen uptake (VO_{2max}) is due to an increase in the central and peripheral components.

$$\mathbf{VO_{2max} = (Q_{max}) \times (a - VO_2diff)_{max}}$$

$$\mathbf{Q_{max} = HR_{max} \times SV_{max}}$$

(Powers & Howley, 2009)

Central Adaptations (Q)

Increase in maximal cardiac output due to an increase in maximal stroke volume (SV). There is no change or slight decrease in maximal HR

(220 – age). In this equation, the variable age cannot be changed. This means that maximal HR changes are dependent on age (see examples below).

Rest

Sedentary individual: 70 (HR) x 71 ml (SV) = 5000 ml (Q)
Trained individual: 50 (HR) x 100 ml (SV) = 5000 ml (Q)

Observe the resting HR and SV during rest.

(Powers & Howley, 2009)

Maximum Exercise

Sedentary individual: 195 (HR) x 113 ml (SV) = 22 000 ml (Q)
Trained individual: 195 (HR) x 179 ml (SV) = 35 000 ml (Q)

Observe the high SV but no change in HR_{max} for both sedentary and trained individuals during exercise.

(Powers & Howley, 2009)

Why Do Some Trained Athletes have SV Values > 210 ml?

The heart (cardiac muscle) possesses similar properties to skeletal muscle. Thus, just like a skeletal muscle, it can also hypertrophy through overload.

- Left ventricular hypertrophy
  - More blood pumped out with each beat due to an increase in left ventricular volume caused by increase in end diastolic volume
- Slower resting HR, more left ventricular filling time
  - Greater volume available for each beat
- Greater elasticity of cardiac muscle fibres
  - More blood volume in the heart's chambers
- Heart: myocardial contractility (increase in myocardial Ca^2 + myosin adenosine triphosphatase (ATPase) activity) (Berchtold, Brinkmeier, & Muntener, 2000; Scott, Stevens, & Binder-Macleod, 2001)

- Greater contractility of cardiac muscles means stronger injection of blood into and out of the chambers

Peripheral Adaptations (a – VO_2diff)

Peripheral cardiovascular adaptations refer to the adaptations that occur in the vasculature or the muscles that contribute to an increased ability to extract oxygen (Powers & Howley, 2009).

$$VO_2 = Q \times (a - VO_2\text{diff})$$

(Powers & Howley, 2009)

Increase in (a – VO_2diff) due to:

- Increase in capillary density
  - More capillaries will provide more perfusion of the muscle
  - Increase in oxygen delivery to muscle
- Increase capacity for maximal vasodilation (diverts more blood)
  - More blood delivered to the muscle
- Increases in the number, size, and membrane surface area of skeletal muscle mitochondria
- Increase in myoglobin content (haemoglobin in the muscle)
  - Increased myoglobin concentration means higher concentration of myoglobin in the muscle. More oxygen is being transported to the tissues.

Metabolic Adaptations

1. Smaller increase in blood and skeletal muscle lactate concentrations in endurance-trained athletes compared to sedentary participants
   - As you become better trained, your blood lactate concentration is lower for the same rate of work. This suggests that you are developing either a greater aerobic capacity, a reduced reliance on the glycolytic system for energy, or perhaps both (Balasekaran,

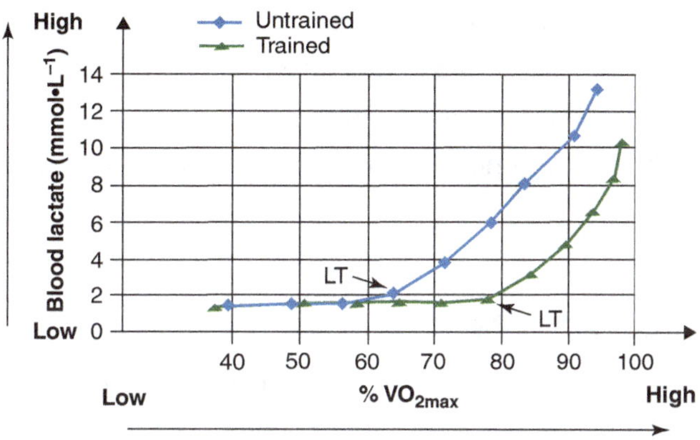

Figure 2. Effects of exercise intensity on blood lactate accumulation in untrained and endurance-trained participants' lactate thresholds (LT) (adapted from Powers & Howley, 2009).

1999; Balasekaran, 2002; Riechman, Zoeller, Balasekaran, Goss, & Robertson, 2002). For an example, for a 200-m swim at a pre-determined speed, an individual has a lactate of approximately 13 $mmol \cdot L^{-1}$ before any training. Due to prolonged training (7 months), his lactate reduces consistently every month until the last month where it is the lowest of about 3 $mmol \cdot L^{-1}$ at the same 200-m swim at a pre-determined speed. This shows greater reliance on aerobic system. This kind of lactate monitoring can be done in any sport. With proper training, there can be a reduced reliance on anaerobic glycolysis and increased utilisation of the aerobic system.

2. Rightward shift of the lactate threshold curve (Figure 2)
   - The trained participant has lactate threshold to the right when compared to the untrained participant (Figure 2), which is due to the trained participant's better cardiovascular and physiological adaptations through prolonged training. Additionally, the rightward shift of the lactate threshold is a cardiovascular and physiological adaptation which happens to the same individual when comparing pre- and post-lactate threshold data after prolonged training (Refer

to *Applied Physiology of Exercise Laboratory Manual* Laboratory Session 10, Balasekaran, Govindaswamy, Lim, Boey, & Ng, 2021, for determination of lactate threshold).

3. Anaerobic training improves muscle buffering capacity (Chapter 3 Figure 3)

Biochemical Adaptations

1. Increase in the level of activity or concentration of the enzymes or increase in the number of enzymatic activity of mitochondria involved in the Krebs cycle and electron transport system (Chapter 4 Figures 2, 3, & 5)
    - An increased level of activity of these enzymes as a result of training means that more adenosine triphosphate (ATP) can be produced in the presence of oxygen (e.g. succinate dehydrogenase, citrate synthase, cytochrome oxidase — enzymes involved in aerobic metabolism) (Chapter 4 Figures 2, 3, & 5)
2. Increase glycogen storage
    - Usually 13 to 15 grams of glycogen per kilogram of muscle. This amount has been shown to increase 2.5 times after training (Powers & Howley, 2009).
3. Decreased utilisation of carbohydrate and increased utilisation of fat during exercise at the same absolute intensity (Chapter 4 Figure 6)
4. The greater work rate (energy expenditure) required to attain the same relative VO_2 following training is met entirely by a proportional increase of fat oxidation. Thus, there is more aerobic contribution than anaerobic contribution.
5. Increased oxidation of carbohydrates (glycogen)
    - Training increases the capacity of skeletal muscle to break down glycogen in the presence of oxygen (oxidise) to CO_2 + H_2O with ATP production (Powers & Howley, 2009) (Chapter 3 Figure 1)
6. Increased oxidation of fat: breakdown of fat to CO_2 + H_2O with ATP production in the presence of oxygen is increased

- An increase in the intramuscular stores of triglycerides, the storage form of fat
- An increase in the release of free fatty acids from adipose tissues, i.e. the availability of fats as fuel is increased
- An increase in the activities involved in the activation, transport, and breakdown of fatty acids
- Increase in the enzymes necessary to break down the large fat molecules into smaller units in preparation for entry into the Krebs cycle and electron transport system (called beta oxidation) (Chapter 4 Figure 6)

(Powers & Howley, 2009)

Specificity of Training

Anaerobic training increases ATP-PC stores, creatine kinase enzyme, glycolytic enzymes (hexokinase, and especially phosphofructokinase (PFK) (most important enzyme in anaerobic glycolysis)) but has no effect on the oxidative enzymes (Chapter 2 Figure 2, Chapter 3 Figure 2). Conversely, aerobic training leads to an increase in the oxidative enzymes, but has little effect on the ATP-PC or glycolytic enzymes (Balasekaran, 1999; Balasekaran, 2002; Riechman, Zoeller, Balasekaran, Goss, & Robertson, 2002).

Anaerobic training improves muscle buffering capacity, but aerobic training does little to enhance the muscles' capacity to tolerate sprint type activities (Chapter 3 Figure 3).

This fact reinforces a recurring theme: **Physiological alterations that result from training are highly specific to the type of training**.

(adapted from Wilmore & Costill, 1999; Plowman & Smith, 2006)

1. **Give examples of specific aerobic training. Name the enzymes which will increase because of the specificity in training. Also give examples of training to increase specific enzymes. (Hint: Chapter 4 Figures 2 & 3)**

2. **What are the types of anaerobic enzymes that increase due to specific training? Give examples of training to increase specific enzymes. (Hint: Chapters 2 & 3)**

3. **Give examples of specific training to improve central adaptations. Give structured and unstructured training. List the specific central adaptations due to your training. (Hint: use the equation $VO_2 = Q \times (a - VO_2diff)$)**

4. **Give examples of specific training to improve peripheral adaptations. Give structured and unstructured training. List the specific peripheral adaptations due to your training. (Hint: use the equation $VO_2 = Q \times (a - VO_2diff)$) (Chapter 2 Figure 2, Chapter 3 Figures 1 & 2, Chapter 4 Figures 2 & 3)**

Chapter 8

Thermoregulation

Terminology

Normohydration: normal body water volume

Euhydration: normal body water volume

Hypohydration: low body water volume

Hyperhydration: excess body water volume

Measurement of Body Temperature

- Core temperature (T_{co})
  - T_{re}: rectal temperature
  - T_{tym}: tympanic membrane temperature
  - T_{eso}: esophageal temperature
- Skin temperature (T_{sk})

Body's core temperature: 36.9°C or 98.6°F

Heat

- Heat can be gained from the environment
- As a result of metabolic activity, the body typically produces majority of the heat
- Heat is a by-product of cellular respiration

The minimum energy required to meet the metabolic demands of the body at rest is called basal metabolic rate or resting metabolic rate (refer to *Applied Physiology of Exercise Laboratory Manual* Laboratory Session 3, Balasekaran, Govindaswamy, Lim, Boey, & Ng, 2021) to calculate resting metabolic rate. It accounts for a large proportion of heat production. At rest, the body liberates 60% to 75% of energy from aerobic metabolism as heat (Powers & Howley, 2009).

The ingestion of food increases the body's production of heat and is known as thermogenesis.

Metabolism is greatly increased during physical activity. Heat production is also increased dramatically.

Heat can be exchanged (gained or lost from the body) by 4 processes (Figure 1):

1) Radiation
2) Conduction
3) Convection
4) Evaporation

(Plowman & Smith, 1997; Powers & Howley, 2009)

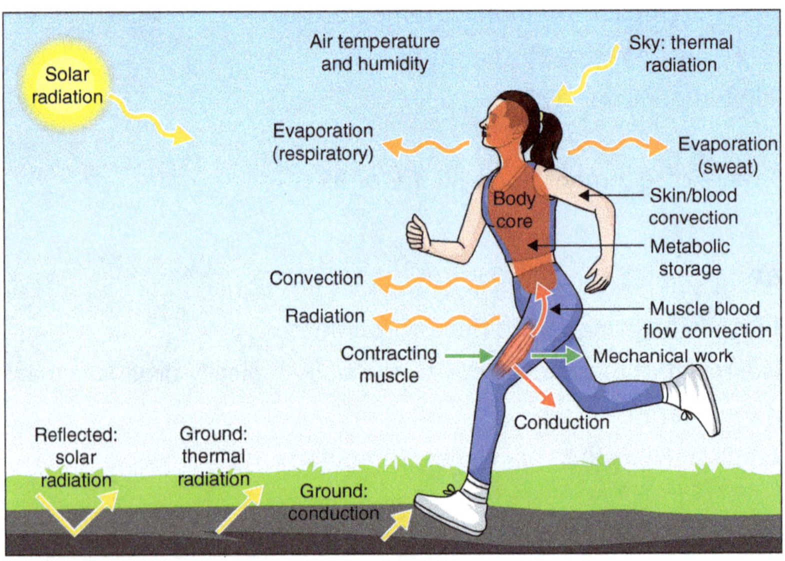

Figure 1. The different processes of heat exchange.

Effectiveness of these processes depends on environmental conditions namely:

- Ambient temperature
- Relative humidity
- Wind speed

(Plowman & Smith, 1997; Powers & Howley, 2009)

1) Radiant Heat Loss

Radiant heat loss occurs through the emission of electromagnetic heat waves from the sun to the environment and depends upon the thermal gradient between the body and the environment.

When environmental temperature = skin temperature, no heat is lost through radiation.

If environmental temperature > skin temperature, radiation will add to the heat load of the body.

2) Conduction

Conduction is the direct transfer of heat from one molecule (direct transfer of heat from one body to another). It occurs through the warming molecules of air and other surfaces in contact with skin.

Water can absorb and conduct heat much better than air. Hence, submersion in water is an effective way to lower body temperature.

3) Convection

Convection is a collected movement of molecules that are in contact with the skin.

When there is a breeze, heat loss is enhanced because the warmer molecules are moved away from the skin.

1. What is the difference between conduction and convection?

4) Evaporation

Evaporation is the conversion of liquid into water vapour.

The evaporation of sweat is a major mechanism for cooling the body under resting conditions. It is dependent on:

- Environmental conditions
- Exercise intensity
- Fitness level
- Degree of acclimatisation
- Hydration status

(Plowman & Smith, 1997; Powers & Howley, 2009)

Evaporation is the primary defense against heat stress. Energy is needed to convert the liquid sweat into vapour. Approximately 580 kcals of heat energy is needed for the evaporation of each litre of sweat (Guyton and Hall, 2011). Sweating itself does not cool the body; it is the evaporation of sweat that cools the body.

Measurement of Environment Conditions

Relative Humidity

- The moisture in the air relative to how much moisture, or water vapour, can be held by the air at any given ambient temperature (Figure 2)

When the body is in thermal balance, the amount of heat produced and the amount of heat lost are equal.

$$\mathbf{M \pm R \pm C \pm K - E = 0}$$

M: metabolic heat production
R: radiant heat exchange
C: convective heat exchange
K: conductive heat exchange
E: evaporative heat loss

(Plowman & Smith, 1997)

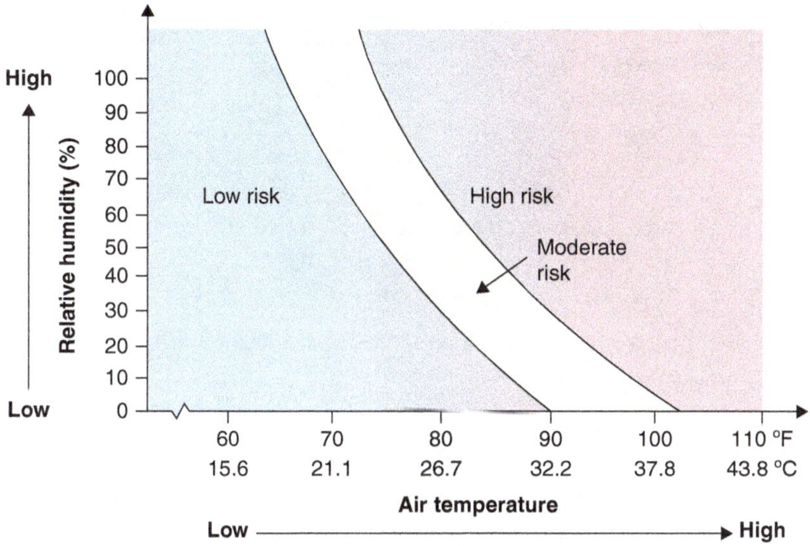

Figure 2. Heat stress index (adapted from Plowman & Smith, 1997).

When the environment is hotter than skin temperature, heat is gained by the body (+).

When skin temperature is higher than environmental temperature, heat is lost from the body (–).

It is not possible for evaporation to add to the heat load of the body. This mechanism can only dissipate heat, thus there is only a negative sign in the equation for evaporation.

There are several factors involved in heat exchange. The effectiveness of heat exchange between an individual and the environment is affected by 5 factors.

1. Thermal gradient
    - The greater the difference between two temperatures (thermal gradient), the greater the heat loss is from the warmer of the two
2. Relative humidity
3. Air movement

4. The degree of direct sunlight
5. The clothing worn by the individual

(Plowman & Smith, 1997)

1. Is it good to run/exercise shirtless on a hot sunny day?

2. Is wearing a black shirt to run/exercise a logical way to thermo-regulate the body?

3. If you run/do an activity on a hot day or under a mid-afternoon sun, what would your running/activity intensity and duration be? Does your intensity and duration depend on your fitness level in such a situation? Give physiological reasons to both questions.

4. Does running/doing an activity in the shade or near a windy environment (e.g. next to the sea, at the beach) help you succeed or complete your running/activity? Give physiological and environmental reasons to both questions.

Exercising in the Heat: Central and Peripheral Demands

$$VO_{2max} = Q_{max} \times (a - VO_2 diff)_{max}$$
$$Q_{max} = HR_{max} \times SV_{max}$$

(Powers & Howley, 2009)

Cardiac output (Q) increases to a similar degree when exercise is performed in a hot or thermoneutral environment.

Q in a hot environment is achieved by a higher heart rate (HR) and a lower stroke volume (SV) than in a thermoneutral environment.

Reduction of Stroke Volume

During hot conditions, there is vasodilation in the blood (cutaneous) vessels, which decreases venous volume. A progressive reduction in SV is observed as the severity of exercise increases. The higher HR seen is not able to compensate for the reduction in SV. Thus, Q is lower in hot conditions.

Vasoconstriction occurs in the digestive and renal areas in an attempt to maintain mean arterial blood pressure. Under severe conditions, vasoconstriction may result in ischemia and even tissue damage.

(Powers & Howley, 2009)

Maximal Oxygen Consumption (VO_{2max})

VO_{2max} is lower in hot environments.

As skin blood flow accounts for the larger portion of the blood flow, more blood is channelled to cool the blood through evaporation. Chapter 5 Figure 3 and Chapter 8 Figure 3 indicate normal skin blood flow during heavy exercise. However, during exercise in a hot environment, skin blood flow will increase (Figure 3). Thus, there will be a reduction in Q. Cooling of the blood will reduce the core temperature, which would have increased as a result of metabolic activity (e.g. high-intensity exercise which produces lactate). Thus, blood flow to the peripheral is compromised as less oxygen is delivered, there is less extraction by the muscle tissue, and ($a - VO_2diff$) is also affected.

The decrease in blood flow and lower cardiac output both contribute to a lower VO_{2max}.

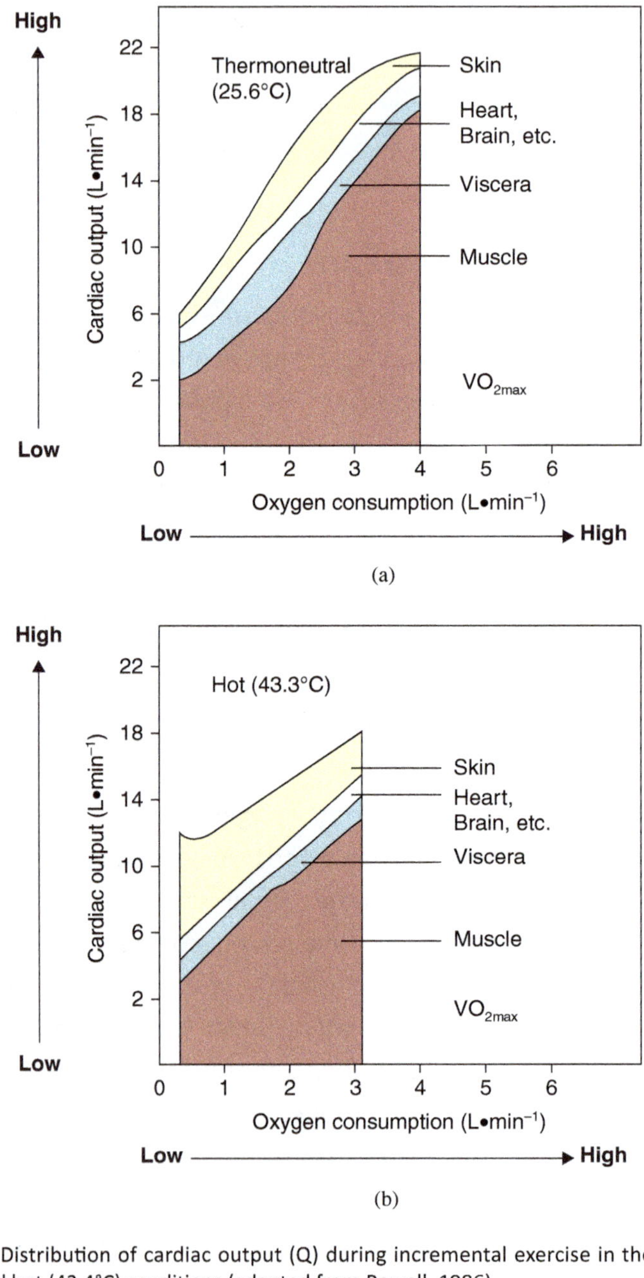

Figure 3. Distribution of cardiac output (Q) during incremental exercise in thermoneutral (25.6°C) and hot (43.4°C) conditions (adapted from Rowell, 1986).

Factors Affecting Cardiovascular Response to Exercise in the Heat

Acclimatisation

The adaptive changes that occur when an individual undergoes prolonged or repeated exposure to a stressful environment. These changes reduce the physiological strain produced by such environments.

- Lowers thermal and cardiovascular strain
- Cardiovascular changes occur that decrease HR and cardiovascular strain at a given level of exercise in the heat
- Sweating patterns are altered such that sweating begins earlier
- The higher sweating rate for a given core temperature can be maintained for longer periods of time

(Plowman & Smith, 1997; Powers & Howley, 2009)

Fitness Level

Aerobic fitness improves an individual's thermoregulatory function and heat tolerance.

- Lower resting core temperature
- Larger plasma volume (refer to *Applied Physiology of Exercise Laboratory Manual* Laboratory Session 13, Balasekaran, Govindaswamy, Lim, Boey, & Ng, 2021)
- Earlier onset of sweating
- Less loss of electrolytes (Figure 4)

Electrolyte	Blood	Sweat of unacclimatised, unfit subject	Sweat of fit but unacclimatised subject	Sweat of fit, acclimatised subject
Sodium (Na^+)	140 ± 6.1	80 ± 3.5	60 ± 2.6	40 ± 1.8
Potassium (K^+)	4.0 ± 0.1	8.0 ± 0.2	6.0 ± 0.15	4.0 ± 0.1
Magnesium (Mg^+)	1.5 ± 0.1	1.5 ± 0.1	1.5 ± 0.1	1.5 ± 0.1
Chloride (Cl^-)	101± 2.9	50 ± 1.4	40 ± 1.1	30 ± 0.9

Note. All values in mmol/L (g/L). Based on data from Verde, T., Shephard, R. J., Corey, P., and Moore, R. (1982), Sweat Composition in Exercise and in Heat, *Journal of Applied Physiology*, *53*(6), pp. 1541, 1543; Costill, D. L. (1997), Sweating: Its Composition and Effects on Body Fluids, *Annals of the New York Academy of Sciences*, 301, p. 162; and Noakes, T. (2003). *Lore of Running.* Human Kinetics.

Figure 4. Electrolyte contents of sweat and blood and the effects of fitness and heat acclimatisation.

Body Composition

Excessive body fat is a liability in terms of thermoregulation during exercise in the heat (Mayo & Balasekaran, 2002; Balasekaran, 2003; Mayo, Grantham, & Balasekaran, 2003; Lee & Balasekaran, 2010; Gupta, Balasekaran, Govindaswamy, Chia, & Lim, 2011; Balasekaran, Gupta, Govindaswamy, Wang, & Bakri, 2014).

- Dissipation of heat
- Insulates the core temperature
- Metabolic cost of activity increases by adding weight to the body

Hydration Level

- Plasma volume decreases

Lower Percentage of Plasma in the Blood

- Acute fluid loss and the inability of the cardiovascular system to adequately compensate for the concurrent demands of muscle and skin blood flow
- Children are more prone to fluid loss as they have lower plasma volume

Heat Illness

What is the Definition of Heat Illness?

1. **Heat cramps**
   - May occur during intense physical activities
   - Fluid electrolyte imbalance: an acute disorder consisting of brief, recurrent, and excruciating pain in the voluntary muscles of the legs, arms, and/or abdomen

2. **Heat syncope (fainting)**

   A temporary disorder characterised by circulatory failure due to the pooling of blood in the peripheral veins and the subsequent decrease in ventricular filling, which leads to a decrease in Q ($VO_2 = Q \times (a - VO_2diff)$, ($Q = HR \times SV$)) (Plowman & Smith, 1997).
   - Individual feels lightheaded and might faint
   - May occur during strenuous and unaccustomed exercise — unable to cope with the sudden rise in temperature or humidity

3. **Heat exhaustion**

Acute fluid loss and the inability of the cardiovascular system to adequately compensate for the concurrent demands of muscle and skin blood flow. Children are more prone to fluid loss as they have a lower plasma volume.

- Rapid and weak pulse
- Fatigue
- Weakness
- Psychological disorientation
- Fainting

4. **Heat stroke**

Failure of the thermoregulatory mechanism

- Serious medical emergency
- Elevated skin and core temperatures
- Rapid HR
- Vomiting
- Diarrhoea
- Hallucinations
- Coma
- Skin is dry, hot, and red (sweating has stopped)

(Plowman & Smith, 1997; Powers & Howley, 2009)

1. As a teacher/coach, what should you do to prevent heat-related illnesses in school/training/exercise?

2. Explain why some people do not sweat after exercising.

3. Does heavy sweat loss or perspiration indicate one's fitness level and does it relate to superior performance? Explain.

4. Can a person who is suffering from heat exhaustion escalate to experience heat stroke? Explain.

5. You are teaching an outdoor physical activity class and someone collapses due to heat. How do you determine whether the person is suffering from heat stroke or heat exhaustion? Supposedly he has heat stroke and you called the ambulance. What action(s) of care will you take for the person who collapsed while waiting for the ambulance?

6. It is hard to survive from a heat stroke. If a person survives from a heat stroke, what are the possible effects (physical and mental)? Explain.

Chapter 9 OMNI Rate of Perceived Exertion

In relation to Chapter 7, rating of perceived exertion (RPE) can be used to self-regulate exercise intensity to allow for cardiovascular adaptations to take place (Balasekaran et al., 2020). RPE is defined as the subjective intensity of effort, strain, discomfort, and/or fatigue that is felt during exercise (Noble & Robertson, 1996). RPE is a subjective correlate of collective physiological responses (heart rate (HR), oxygen consumption, lactic acid, pulmonary ventilation, as well as mechanical workload during incremental exercise) which change according to exercise intensities (Plowman & Smith, 2013). The introduction of the internationally recognised OMNI Scale of Perceived Exertion (for children and adults) allows individuals to exercise within safe intensity levels by assessing RPE responses during and immediately after physical activity participation (Robertson et al., 2000, 2001, 2004; Utter et al., 2002; Balasekaran, Loh, Govindaswamy, & Robertson, 2012; Balasekaran, Loh, Govindaswamy, & Cai, 2014; Balasekaran, Thor, Ng, & Govindaswamy, 2014; Balasekaran, Ismail, & Thor, 2015; Balasekaran et. al., 2019; Balasekaran, Govindaswamy, Ng, & Boey, 2019; Balasekaran, Boey, & Ng, 2018, 2020), thus keeping exercise intensities at prescribed levels.

The Children's OMNI Scale of Perceived Exertion was developed by Robertson et al. in 2000. Its pictorial format provides a pragmatic way to track exercise intensity levels (Robertson et al., 2000, 2001, 2004; Utter, Robertson, Nieman, & Kang, 2002), providing a useful and effective tool to monitor each child's effort during physical activity. This scale has been validated across age groups, genders, and fitness levels of both children (Cai & Balasekaran, 2004; Loh & Balasekaran, 2004; Balasekaran, Loh,

Govindaswamy, & Robertson, 2012; Thor & Balasekaran, 2012; Balasekaran, Thor, Ng, & Govindaswamy, 2014; Chia & Balasekaran, 2018) and adults (Goss et al., 2011; Balasekaran, Boey, & Ng, 2018, 2020). The OMNI RPE scale has been validated for self-regulation of exercise intensity where children are able to produce their estimated intensity accurately (Thor & Balasekaran, 2012).

The OMNI RPE scale is available in formats for both children and adults participating in aerobic and resistance exercises. This encourages young users to progress with their physical activities and to exercise in a safe and effective manner throughout their life. Physical education conducted at inappropriate intensity levels may lead to consequences like excessive fatigue and increase levels of reluctance towards physical activity. Appropriate exercise intensity is vital to enable students to enjoy the benefits and pleasure of an active lifestyle. With the availability of aerobic and resistance exercise scale formats, users can be assured of a balanced improvement in both the cardiovascular and musculoskeletal systems.

Both OMNI RPE scales (for children and adults) can be obtained from the Curricular Guide to Health-Fitness Applications in Physical Education Using the OMNI Perceived Exertion Scale developed by the authors (Balasekaran et al., 2019).

Below are a few examples of how the OMNI RPE scale can be included in a 30-minute (Table 1) and 45-minute (Table 2) exercise for children or adolescents. Likewise, individuals with an objective to lose weight, maintain a healthy lifestyle, or improve their physical fitness may refer

Table 1. Example of an ideal 30-minute lesson with corresponding target OMNI RPE.

Time (minutes)	**5**	**20**	**5**
Activities	Warm up (e.g. slow jog, stretch)	Exercise/game play (e.g. invasion games like hockey, soccer, captain's ball, etc.)	Cool down (e.g. slow jog, stretch)
Intensity	Low	Moderate	Low
RPE (OMNI)	0, 1, 2, 3	4, 5, 6	0, 1, 2, 3

Table 2. Example of an ideal 45-minute lesson with corresponding target OMNI RPE.

Time (minutes)	**5**	**15**	**5**	**15**	**5**
Activities	Warm-up (e.g. slow jog, stretch)	Exercise/game play (e.g. invasion games like hockey, soccer, captain's ball, etc.)	Rest	Exercise/ game play (e.g. invasion games like hockey, soccer, captain's ball, etc.)	Cool down (e.g. slow jog, stretch)
Intensity	Low	Moderate		Moderate	Low
Time (minutes)	5	15		15	5
RPE (OMNI)	0, 1, 2, 3	4, 5, 6		4, 5, 6	0, 1, 2, 3

to the following programs. Examples of 30-minute and 45-minute lesson plans are shown in Tables 1 and 2, respectively.

In addition, HR monitors (chest and wrist) are also a useful tool to monitor the exercise intensity of each individual. However, it is optional as it may be costly, and the OMNI RPE scale can be used without compromising accuracy and reliability of results (Balasekaran et al., 2020). The OMNI RPE scale can be used for individual and group work. It is more useful for teachers and exercise professionals who are often involved in monitoring intensity in group work. Exercising at a moderate intensity (OMNI RPE 4 to 6 for children and OMNI RPE 5 to 7 for adults) improves aerobic fitness at an optimal rate and minimizes injuries. Hence, making use of the OMNI RPE scale during physical activities, physical education lessons, or training allow teachers/coaches/trainers to track and have better control over the intensity of their students' fitness activities and to achieve the desired physiological training responses (refer to *Applied Physiology of Exercise Laboratory Manual* Laboratory Sessions 5 & 6, Balasekaran, Govindaswamy, Lim, Boey, & Ng, 2021).

1. **As a teacher/practitioner/coach, how would you use the OMNI RPE scale and what intensity would you prescribe during your lesson/training/physical activity? Elaborate with physiological reasons for exercising at this intensity.**

2. **Why is there a need for discrimination of intensity for warm up, main activity, and cool down? Explain.**

3. **When will you prescribe exercise intensity at OMNI RPE 8, 9, and/or 10? Explain and justify the period of training using the principles of periodisation. How long should the athlete go through the conditioning phase prior to this training?**

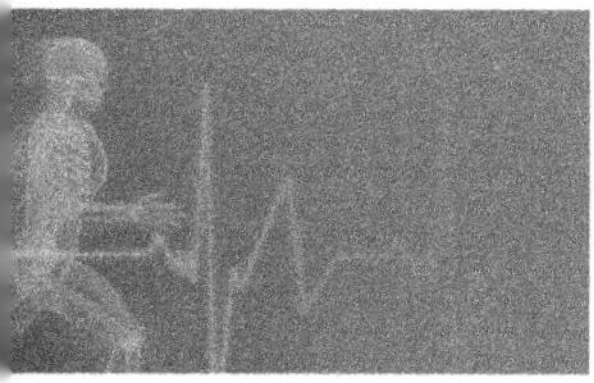

Appendix A: PACER Test/Multistage Fitness Test Lap Check Chart

PACER Test/Multistage Fitness Test Lap Check Chart

Name: ______________________________

Date: ________________________________ **Time:** ____________ **am/pm**

Temperature: _________________________ **Place:** ____________

Instructions: Draw a cross in the box as each lap is completed

Stage	Laps														
	1	2	3	4	5	6	7	8	9						
5	1	2	3	4	5	6	7	8	9						
6	1	2	3	4	5	6	7	8	9	10					
7	1	2	3	4	5	6	7	8	9	10					
8	1	2	3	4	5	6	7	8	9	10					
9	1	2	3	4	5	6	7	8	9	10	11				
10	1	2	3	4	5	6	7	8	9	10	11				
11	1	2	3	4	5	6	7	8	9	10	11	12			
12	1	2	3	4	5	6	7	8	9	10	11	12			
13	1	2	3	4	5	6	7	8	9	10	11	12	13		
14	1	2	3	4	5	6	7	8	9	10	11	12	13		
15	1	2	3	4	5	6	7	8	9	10	11	12	13		
16	1	2	3	4	5	6	7	8	9	10	11	12	13	14	
17	1	2	3	4	5	6	7	8	9	10	11	12	13	14	
18	1	2	3	4	5	6	7	8	9	10	11	12	13	14	15
19	1	2	3	4	5	6	7	8	9	10	11	12	13	14	15
20	1	2	3	4	5	6	7	8	9	10	11	12	13	14	15

Final Score: **STAGE** [ ] **LAP** [ ]

Predicted VO_{2max}: [ ]

Appendix B: PACER Test/Multistage Test Time Elapse Chart (Running Speed (km/h) and VO_{2max} (mL/km/min) Chart)

Test Time Elapse Chart

LEVEL	SHUTTLE	BEEP 1#	ESTIMATED RUNNING SPEED (KM/HR)	ESTIMATED VO_2 MAX (ML/KM/MIN)
4	1	24	10.00	26.4
	2	25	10.05	26.8
	3	26	10.11	27.2
	4	27	10.16	27.6
	5	28	10.22	27.9
	6	29	10.28	28.3
	7	30	10.33	28.7
	8	31	10.39	29.1
	9	32	10.44	29.5
5	1	33	10.50	29.8
	2	34	10.55	30.2
	3	35	10.61	30.6
	4	36	10.66	31.0
	5	37	10.72	31.4
	6	38	10.78	31.8
	7	39	10.83	32.2
	8	40	10.89	32.6
	9	41	10.94	33.9
6	1	42	11.00	33.3
	2	43	11.05	33.6
	3	44	11.10	33.9
	4	45	11.15	34.3
	5	46	11.20	34.6
	6	47	11.25	35.0
	7	48	11.30	35.3
	8	49	11.35	35.7
	9	50	11.40	36.0
	10	51	11.45	36.4
7	1	52	11.50	36.7
	2	53	11.55	37.1
	3	54	11.60	37.4
	4	55	11.65	37.8
	5	56	11.70	38.1
	6	57	11.75	38.5
	7	58	11.80	38.8
	8	59	11.85	39.2
	9	60	11.90	39.5
	10	61	11.95	39.9
8	1	62	12.00	40.2
	2	63	12.05	40.5
	3	64	12.09	40.8
	4	65	12.14	41.1
	5	66	12.18	41.4
	6	67	12.23	41.8
	7	68	12.27	42.1
	8	69	12.32	42.4
	9	70	12.36	42.7
	10	71	12.41	43.0
	11	72	12.45	43.3
9	1	73	12.50	43.6
	2	74	12.55	43.9
	3	75	12.59	44.2
	4	76	12.64	44.5
	5	77	12.68	44.8
	6	78	12.73	45.2
	7	79	12.77	45.5
	8	80	12.82	45.8
	9	81	12.86	46.1
	10	82	12.91	46.4
	11	83	12.95	46.8

LEVEL	SHUTTLE	BEEP 1#	ESTIMATED RUNNING SPEED (KM/HR)	ESTIMATED VO_2 MAX (ML/KM/MIN)
10	1	84	13.00	47.1
	2	85	13.05	47.4
	3	86	13.09	47.7
	4	87	13.14	48.0
	5	88	13.18	48.3
	6	89	13.23	48.7
	7	90	13.27	49.0
	8	91	13.32	49.3
	9	92	13.36	49.6
	10	93	13.41	49.9
	11	94	13.45	50.2
11	1	95	13.50	50.5
	2	96	13.54	50.8
	3	97	13.58	51.1
	4	98	13.62	51.4
	5	99	13.67	51.6
	6	100	13.71	51.9
	7	101	13.75	52.2
	8	102	13.79	52.5
	9	103	13.83	52.8
	10	104	13.87	53.1
	11	105	13.92	53.4
	12	106	13.96	53.7
12	1	107	14.00	54.0
	2	108	14.04	54.3
	3	109	14.08	54.5
	4	110	14.12	54.8
	5	111	14.17	55.1
	6	112	14.21	55.4
	7	113	14.25	55.7
	8	114	14.29	56.0
	9	115	14.33	56.2
	10	116	14.37	56.5
	11	117	14.42	56.8
	12	118	14.46	57.1
13	1	119	14.50	57.3
	2	120	14.54	57.6
	3	121	14.58	57.9
	4	122	14.61	58.2
	5	123	14.65	58.4
	6	124	14.69	58.7
	7	125	14.73	59.0
	8	126	14.77	59.3
	9	127	14.81	59.5
	10	128	14.85	59.8
	11	129	14.88	60.0
	12	130	14.92	60.3
	13	131	14.96	60.6
14	1	132	15.00	60.8
	2	133	15.04	61.1
	3	134	15.08	61.4
	4	135	15.11	61.7
	5	136	15.15	61.9
	6	137	15.19	62.2
	7	138	15.23	62.4
	8	139	15.27	62.7
	9	140	15.31	62.9
	10	141	15.35	63.2
	11	142	15.38	63.4
	12	143	15.42	63.7
	13	144	15.46	64.0

LEVEL	SHUTTLE	BEEP 1#	ESTIMATED RUNNING SPEED (KM/HR)	ESTIMATED VO_2 MAX (ML/KM/MIN)
15	1	145	15.50	64.3
	2	146	15.54	64.6
	3	147	15.58	64.8
	4	148	15.61	65.1
	5	149	15.65	65.3
	6	150	15.69	65.6
	7	151	15.73	65.9
	8	152	15.77	66.2
	9	153	15.81	66.4
	10	154	15.85	66.7
	11	155	15.88	66.9
	12	156	15.92	67.2
	13	157	15.96	67.5
16	1	158	16.00	67.7
	2	159	16.04	68.0
	3	160	16.07	68.2
	4	161	16.11	68.5
	5	162	16.14	68.7
	6	163	16.18	69.0
	7	164	16.21	69.2
	8	165	16.25	69.5
	9	166	16.29	69.7
	10	167	16.32	69.9
	11	168	16.36	70.2
	12	169	16.39	70.5
	13	170	16.43	70.7
	14	171	16.46	70.9
17	1	172	16.50	71.1
	2	173	16.54	71.4
	3	174	16.57	71.6
	4	175	16.61	71.9
	5	176	16.64	72.1
	6	177	16.68	72.4
	7	178	16.71	72.6
	8	179	16.75	72.9
	9	180	16.79	73.1
	10	181	16.82	73.4
	11	182	16.86	73.6
	12	183	16.89	73.9
	13	184	16.93	74.1
	14	185	16.96	74.4
18	1	186	17.00	74.6
	2	187	17.03	74.8
	3	188	17.07	75.0
	4	189	17.10	75.3
	5	190	17.13	75.5
	6	191	17.17	75.8
	7	192	17.20	76.0
	8	193	17.23	76.2
	9	194	17.27	76.4
	10	195	17.30	76.7
	11	196	17.33	76.9
	12	197	17.37	77.2
	13	198	17.40	77.4
	14	199	17.43	77.6
	15	200	17.47	77.9

Note: Speed can be used for training purposes for track or road with fixed distances (e.g. interval training)

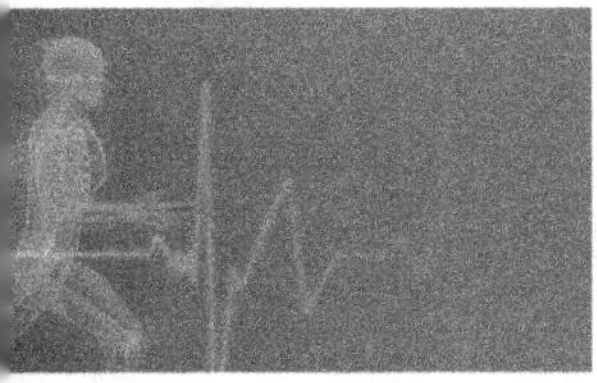

Appendix C: PACER Test/Multistage Oxygen Uptake Values

TABLE OF PREDICTED MAXIMUM OXYGEN UPTAKE VALUES

Adapted from the Multistage Fitness Test
Department of Physical Education and Sports Science
Loughborough University, 1987

LEVEL	SHUTTLE	PREDICTED VO_{2MAX} $(mL \cdot kg^{-1} \cdot min^{-1})$
4	2	26.8
4	4	27.6
4	6	28.3
4	9	29.5
5	2	30.2
5	4	31.0
5	6	31.8
5	9	32.9
6	2	33.6
6	4	34.3
6	6	35.0
6	8	35.7
6	10	36.4
7	2	37.1
7	4	37.8
7	6	38.5
7	8	39.2
7	10	39.9
8	2	40.5
8	4	41.1
8	6	41.8
8	8	42.4
8	10	43.3

LEVEL	SHUTTLE	PREDICTED VO_{2MAX} $(mL \cdot kg^{-1} \cdot min^{-1})$
9	2	43.9
9	4	44.5
9	6	45.2
9	8	45.8
9	11	46.8
10	2	47.4
10	4	48.0
10	6	48.7
10	8	49.3
10	11	50.2
11	2	50.8
11	4	51.4
11	6	51.9
11	8	52.5
11	10	53.1
11	12	53.7
12	2	54.3
12	4	54.8
12	6	55.4
12	8	56.0
12	10	56.5
12	12	57.1

LEVEL	SHUTTLE	PREDICTED VO_{2MAX} ($mL \cdot kg^{-1} \cdot min^{-1}$)	LEVEL	SHUTTLE	PREDICTED VO_{2MAX} ($mL \cdot kg^{-1} \cdot min^{-1}$)
13	2	57.5	17	2	71.4
13	4	58.1	17	4	71.9
13	6	58.8	17	6	72.4
13	8	59.4	17	8	72.0
13	10	59.3	17	10	73.4
13	13	60.3	17	12	73.0
			17	14	74.4
14	2	61.1	18	2	74.8
14	4	61.7	18	4	75.3
14	6	62.2	18	6	75.8
14	8	62.7	18	8	76.2
14	10	63.2	18	10	76.7
14	13	64.0	18	12	77.2
			18	13	77.0
15	2	64.5	19	2	78.3
15	4	65.1	19	4	78.8
15	6	66.8	19	6	79.2
15	8	66.4	19	8	79.7
15	10	66.3	19	10	80.2
15	13	67.3	19	12	80.6
			19	13	81.3
16	2	68.0	20	2	81.8
16	4	68.5	20	4	82.2
16	6	69.0	20	6	82.6
16	8	69.5	20	8	83.0
16	10	69.9	20	10	83.5
16	12	70.5	20	12	83.0
16	14	70.9	20	13	84.8

Appendix D: One-Mile Walk Test for VO_{2max}

This cross-validated test provides a non-traumatic estimate of maximal oxygen consumption (L/min) for healthy individuals of both genders, and is especially suitable for children, obese individuals, and those with orthopaedic problems. Its validity is probably enhanced by the inclusion of factors such as age, weight, gender, exercise heart rate, and of course, one-mile time. The purpose of the one-mile walk test is to measure aerobic power (VO_{2max} in L/min) by using a regression equation (Kline et al., 1987).

Method

1. Walk as fast as possible for 1 mile; record the time (min:s) and convert to decimal minutes.
   1-mile time
   Decimal min (t) = [min + (s/6)] = ________
2. Take heart rate (15 s) immediately on crossing the one-mile mark.
3. Multiply the heart rate by 4 in order to convert to beats per minute.
   Heart rate in 15 s _______ × 4 = b/min
4. Use the regression equation to estimate maximal oxygen consumption.

$$VO_{2max}\ (\text{L/min}) = 6.9552 + (0.0091 \times Wt) - (0.0257 \times Age) + (0.5955 \times gender) - (0.2240 \times t) - (0.0115 \times HR)$$

Where: Wt = weight (lb); gender: 1 = male; 0 = female
t = time (decimal min); HR = heart rate (b/min); Age = closest year (y)

Discussion

The validity correlation for this generalised equation using the time for the first one-mile walk versus the directly measured maximal oxygen consumption was high (r = 0.93) and the standard error of estimate (SEE) was 0.326 L/min (Kline et al., 1987). However, these values were based on electronically-monitored pulse rates during field testing. Thus, accurate pulse taking should be emphasised in order to obtain similar valid results.

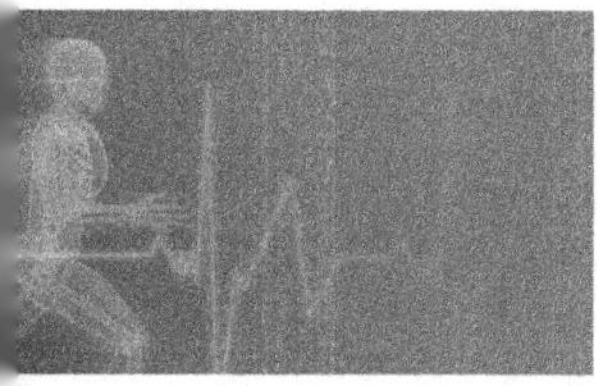

References

1. Ali, M. J., Balasekaran, G., Hoon, K. H., and Gerald, S. (2017). Physiological Differences Between a Non-Continuous and a Continuous Endurance Training Protocol in Recreational Runners and Metabolic Demand Prediction. *Physiological Reports, 1*(1), 30–35.
2. American College of Sports Medicine (Ed.). (2018). *Guidelines for Exercise Testing and Prescription*. Williams & Wilkins.
3. Andreacci, J. L., Robertson, R. J., Dube, J. J., Aaron, D. J., Balasekaran, G., and Arslanian, S. A. (2004). Comparison of maximal oxygen consumption between black and white prepubertal and pubertal children. *Pediatric Research, 56*(5), 706–713.
4. Armstrong, N., and Welsman, J. (1997) *Young People and Physical Activity*. Oxford University Press.
5. Balasekaran, G. (1993). To Determine the Exercise Economy of an Experimental Innersole as Compared to that of a Traditional Innersole while Walking, Jogging and Running on a Treadmill. (Masters dissertation, Indiana University of Pennsylvania, United States).
6. Balasekaran, G. (1999). Determination of a Mass Exponent for Maximal Aerobic Power and Peak Anaerobic Power in African-American and Caucasian Children. (PhD dissertation, University of Pittsburgh, United States).
7. Balasekaran, G. (2001). Physiological Factors of Long Distance Running Affected by Biomechanical Efficiency of Human Movement. *International Association of Athletics Federations Bulletin, 1*, 25–35.

8. Balasekaran, G. (2002). The Physiology of Lactate and its Application in Long Distance Running. *International Athletic Association Federation Bulletin*, (2), 25–30.
9. Balasekaran, G. (2003). Body Composition in Track and Field: Measurement and Application. *International Athletic Association Federation Bulletin, 1*, 30–35.
10. Balasekaran, G. and Loh, M. K. (2009). School Physical Education Programmes: Health & Fitness Issues and Challenges. In Aplin, N. (Ed.), *An Eye on the Youth Olympic Games 2010: Perspectives on PE and Sport Science in Singapore* (pp. 50–60). McGraw-Hill.
11. Balasekaran, G., Boey, P., and Ng, Y. C. (2018). Effects of Self-regulating Exercise Intensity using the OMNI Rate of Perceived Exertion Scale on Youths and Pedagogical Methods for Youths during Physical Education in Singapore. In Popoviæ, S., Antala, B., Bjelica, D., and Gardaševiæ, J. (Eds.), *Physical Education in Secondary School: Researches, Best Practices, Situation* (pp. 129–143). Bratislava: Faculty of Sport and Physical Education of University of Montenegro, Montenegrin Sports Academy and Fédération Internationale D´Éducation Physique (FIEP Europe).
12. Balasekaran, G., Boey, P., and Ng, Y. C. (2020). Best Practices in Physical Education and Physical Activity in Nanyang Technological University Singapore. In Bobrík, M., Antala, B., and Pělucha R. (Eds.), *Physical Education in Universities: Researches, Best Practices, Situation* (pp. 285–293). Bratislava: Slovak Scientific Society for Physical Education and Sport and FIEP.
13. Balasekaran, G., Boey, P., Hui, S. S-C., Govindaswamy, V. V., Ng, Y. C., and Lim, Z. J (2018). Correlation of Handgrip Strength and Cardiovascular Fitness with Percent Body Fat in Singapore Adolescents. *Gazzetta Medica Italiana Archivio per le Scienze Mediche, 177*(5), 198–203.
14. Balasekaran, G., Govindaswamy, V., Ng, Y. C., and Boey, P. (2019). Best Practices in Physical Education in Singapore's Early Childhood Education and Care. In Antala, B., Demirhan, G., Carraro, A., Oktar, Oz, H., and

Kaplanova, A. (Eds.), *Physical Education in Early Childhood Education and Care: Researches, Best Practices, Situation* (pp. 215–25). Bratislava: Slovak Scientific Society for Physical Education and Sport and FIEP.

15. Balasekaran, G., Gupta, N., and V. Govindaswamy, V. V. (2010). Assessment Of Body Composition. In Chia, M. and Chiang, J. (Eds.), *Sport Science & Studies in the East: Issues, Reflections and Emergent Solutions* (pp. 200–220). World Scientific.
16. Balasekaran, G., Gupta, N., Govindaswamy, V. V., Wang, P. K., and Bakri, A. Z. (2014). Physical Education Story: A Journey of Transformations in Singapore. In Chin, M. K. and Edginton, G. (Eds.), *Physical Education and Health: Global Perspectives and Best Practice* (pp. 409–420). Sagamore Publishing LLC.
17. Balasekaran, G., Hui, S. S-C., Thor, D., Govindaswamy, V. V., and Ng, Y. C. (2016). Cardiovascular Fitness, Flexibility and Body Composition Association among Adolescents in Singapore. *Juntendo Journal of Health and Sports Science*, 7(2), 74–81.
18. Balasekaran, G., Ismail, I., and Thor, D. (2015). Effect of 4 weeks of Fun Game-Based and Structured Interval Training Physical Education Lessons on Aerobic Fitness in Adolescents. *Asian Journal of Physical Education and Sport Science, 5*, 1–9.
19. Balasekaran, G., Lim, J. Z., Boey, P., Ng, Y. C., and Govindaswamy, V. V. (2020). AQUATITAN™ lower body compression garment results in lower 200-m run timings. *Gazzetta Medica Italiana Archivio per le Scienze Mediche, 179*(6), 412–418.
20. Balasekaran, G., Lim, Z. J., Boey, P., Govindaswamy, V. V., Foo, W., and Ng, Y. C. (2019). Acute Foam Rolling on Quadriceps Performance and Short-Term Recovery from Fatigue. *Gazzetta Medica Italiana Archivio per le Scienze Mediche, 1*(1), 200–229. (RPE)
21. Balasekaran, G., Lim, Z. J., Govindaswamy, V. V., Ee, S., W., and Ng, Y. C. (2019). Effect of Aquatitan Bracelet on Quadriceps Recovery after Fatiguing Muscular Strength and *Endurance Exercise. Gazzetta Medica Italiana Archivio per le Scienze Mediche, 1*(2), 100–120.

22. Balasekaran, G., Loh M. K., Govindaswamy, V. V., and Cai S. J. (2014). OMNI Scale Perceived Exertion Responses in Obese and Normal Weight Male Adolescents During Cycle Exercise. *Journal of Sports Medicine and Physical Fitness, 54*(2), 186–196.
23. Balasekaran, G., Loh, M. K., Govindaswamy, V. V., and Robertson, R. J. (2012). OMNI Scale of Perceived Exertion: Mixed Gender and Race Validation for Singapore Children During Cycle Exercise. *European Journal of Applied Physiology, 1*, 35–38.
24. Balasekaran, G., Mayo, M., and Lim, J. (2019). Fat Distribution and Metabolic Risk Factors of Young Obese Males Following the Cessation of Training: A Follow Up. *Translational Sports Medicine, 2*(2), 82–89.
25. Balasekaran, G., Roberston, R. J., Goss, F. L., Lewy, V., Danadian, K., and Arslanian, S. A. (2000). Determination of Mass Exponents for Maximal Aerobic Power in African American and Caucasian Children. *Medicine and Science in Sport and Exercise*, 32(5).
26. Balasekaran, G., Roberston, R. J., Goss, F. L., Suprasongsin, C., Dandian, K., and Arslanian, S. A. (1998). Effect of Growth Hormone Therapy on VO2max, Body Composition and Physical Development in Adolescents. *Medicine and Science in Sport and Exercise*, 30(5), 259.
27. Balasekaran G., Roberston, R. J., Riechman, S. E., Goss, F. L., Lewy, V., Danadian, K., and Arslanian, S. A. (2001). Effect of Dihydrotestosterone Therapy on VO2max, Body Composition and Physical development in Adolescents. *Medicine and Science in Sport and Exercise*, 33(5).
28. Balasekaran G., Robertson, R. J., Goss, F. L., and Arslanian, S. A. (2002). Determination of Mass Exponent for Anaerobic Power in African American and Caucasian Children. *Medicine and Science in Sport and Exercise*, 34(5).
29. Balasekaran, G., Robertson, R. J., Goss, F. L., Suprasongsin, C., Danadian, K., Govindaswamy V. V., and Arslanian, S. A. (2005). Short-term pharmacological induced growth study of ontogenetic allometry of oxygen uptake in children. *Annals of Human Biology, 32*(6), 746–759.

30. Balasekaran, G., Thor, D., Ng, Y. C., and Govindaswamy, V. V. (2014). OMNI Scale of Perceived Exertion: Self-Regulation of Exercise Intensity in Youths and Pedagogical Approaches for Youths in Physical Education in Singapore. *Asian Journal of Youth Sport, 1*(1), 43–50.
31. Bar-Or, O. (1983). *Pediatric Sports Medicine*. Springer-Verlag.
32. Bar-Or, O. (1996). *The Child & Adolescent Athlete*. Human Kinetics Publishers .
33. Berchtold, M. W., Brinkmeier, H., & Muntener, M. (2000). Calcium ion in skeletal muscle: its crucial role for muscle function, plasticity, and disease. *Physiological Reviews,* 80(3), 1215–1265.
34. Brown, S. P., Miller, W. C., & Eason, J. M. (2006). *Exercise Physiology: Basis of Human Movement in Health and Disease*. Lippincott Williams & Wilkins.
35. Cai, S. J. and Balasekaran, G. (2004). Validation of OMNI Rate of Perceived Exertion Scale in Obese Male Individuals. (Honours dissertation, National Institute of Education, Nanyang Technological University, Singapore).
36. Caplan, G. (2007). *BTEC National Sport*. Heinemann.
37. Chia, J., and Balasekaran, G. (2018). Rate of Perceived Exertion (RPE) — A Safe Self-Regulation of Exercise Intensity for Children in Singapore. (Masters dissertation, National Institute of Education, Nanyang Technological University, Singapore).
38. Conley, D. L., and Krahenbuhl, G. S. (1980). Running Economy and Distance Running Performance of Highly Trained Athletes. *Medicine and Science in Sports and Exercise, 12*(5).
39. Cooper, K. H. (1968). A Means of Assessing Maximal Oxygen Uptake. *Journal of the American Medical Association, 203*, 201–204.
40. Costill, D. L. (1986). *Inside Running: Basics of Sports Physiology*. Cooper Pub Group.
41. Dabayebeh I., Riechman, S. E., Balasekaran, G., Goss, F. L., Moyna, N., Metz, K., Baker, C., and Robertson, R. J. (1999). Effect of Aerobic Exercise

at Various Sub-Maximal Intensities on Circulating Beta-Endorphin Concentration. *Medicine and Science in Sport and Exercise, 31*(5) 268.

42. Foss, M. L. and Keteyian, S. J. (1998). *Physiological Basis for Exercise and Sport*. McGraw-Hill.
43. Fox, E. L., Bowers, R. W., and Foss, M. L. (1993), *The Physiological Basis for Exercise and Sport*. Brown and Benchmark.
44. Goodwin, M. L., Harris, J. E., Hernández, A., and Gladden, L. B. (2007). Blood Lactate Measurements and Analysis During Exercise: A Guide for Clinicians. *Journal of Diabetes Science and Technology, 1*(4), 558–569.
45. Govindasamy, B., Govindaswamy, V. V., Loh R. M. K., Ng, Y. C., Thor, D., Boey. P. and Lim, J. (2019). *Curricular Guide to Health- Fitness Applications in Physical Education Using the OMNI Perceived Exertion Scale*. Pearson Education South Asia Pte Ltd.
46. Grantham, J. R. and Balasekaran, G. (2004). Effects of Aerobic and Strength Training on Basal Metabolic Rate and Postprandial Lipaemia (PhD dissertation, National Institute of Education, Nanyang Technological University, Singapore).
47. Grantham J. R., Mayo M. J., and Balasekaran, G. (2003). The Effect of 9-weeks Strength Training on Postprandial Lipemia. *Medicine and Science in Sport and Exercise, 35*(5).
48. Gupta, N. and Balasekaran, G. (2013). Running Energy Reserve Index: Mapping, Assessment and Prediction (PhD dissertation, National Institute of Education, Nanyang Technological University, Singapore).
49. Gupta, N., Balasekaran, G., Govindaswamy, V. V., Chia, Y. H., and Lim, M. S. (2011). Comparison of body composition with bioelectric impedance (BIA) and dual energy X-ray absorptiometry (DEXA) among Singapore Chinese. *Journal of Science and Medicine and Sport, 14*(1), 33–35.
50. Hall, J. E., & Guyton, A. C. (2011). *Textbook of Medical Physiology.* Saunders.
51. Holloszy, J. O. (1982). Muscle metabolism during exercise. *Archives of Physical Medicine and Rehabilitation, 63*(5), 231–234.

52. Janssen, P. G. J. M. (1994). *Training Lactate Pulse Rate* (4th ed.). Polar Electro.
53. Kline G. M., Parcari, J. P., Hinermeister, R., Freedson, P. S., Ward, A., McCarron, R. F., Ross, J., and Rippe J. M. (1987). Estimation of $\dot{V}O2max$ from a One-Mile Track Walk, Gender, Age, and Body Weight. *Medicine and Science in Sports and Exercise, 19*(3), 253–259.
54. Lamb, D. R. (1984). *Exercise Physiology*. Macmillan Publishers.
55. Lamb, D. R. (1984). *Physiology of Exercise: Responses and Adaptations*. Macmillan Pub Co.
56. Lee, M. K. and Balasekaran, G. (2010). The Measurement of Three-Dimensional Body Segment Parameters Using Dual Energy X-Ray Absorptiometry and Skin Geometry an Application in Gait Analysis. (PhD dissertation, National Institute of Education, Nanyang Technological University, Singapore).
57. Loh, M. K. and Balasekaran, G. (2004). Validation of Gender and Racially Specific OMNI Perceived Exertion Rating Scales for Children in Singapore (Masters dissertation, National Institute of Education, Nanyang Technological University, Singapore).
58. Mayo, M. J., Grantham, J. R., and Balasekaran, G. (2002). Effects of Large Exercise-Induced Weight Loss on Abdominal Fat and Lipid Levels in Obese Males. *Medicine & Science in Sports & Exercise, 35*(S5), 135.
59. Mayo M. J., Grantham J. R., and Balasekaran G. (2003). Exercise-Induced Weight Loss Preferentially Reduces Abdominal Fat. *Medicine and Science in Sport and Exercise, 35*(2), 207–213.
60. McArdle, W. D., Katch, F. I., and Katch, V. L. (1991). *Exercise Physiology: Energy, Nutrition and Human Performance*. Lippincott Williams and Wilkins.
61. McArdle, W. D., Katch, F. I., and Katch, V. L. (2006). *Exercise Physiology: Nutrition, Energy, and Human Performance*. Lippincott Williams & Wilkins.
62. McArdle, W. D., Katch, F. I., and Katch, V. L. (2007). Exercise Physiology: Nutrition, Energy, and Human Performance. Lippincott Williams & Wilkins.

63. McArdle, W. D., Katch, F. I., and Katch, V. L. (2010). Exercise Physiology: Nutrition, Energy, and Human Performance. Lippincott Williams & Wilkins.
64. Newsholme, E., Leech, T., and Duester, G. (1994). *Keep on Running: The Science of Training and Performance*. Wiley and Sons.
65. Noble, B. J. and Robertson, R. J. (1996). The Borg Scale: Development, Administration, and Experimental Use. Perceived Exertion. Human Kinetics Publishers, 59–92.
66. Plowman, S. A. and Smith D. L. (1997). *Exercise Physiology for Health, Fitness, and Performance*. Allyn and Bacon.
67. Plowman, S. A. and Smith, D. L. (2003). *Exercise Physiology for Health, Fitness, and Performance, 2nd Edition*.
68. Plowman, S. A., and Smith, D. L. (2013). *Exercise Physiology for Health Fitness and Performance*. Lippincott Williams & Wilkins.
69. Pollock, M. L., Wilmore, J. H., and Fox III, S. M. (1978). *Health and Fitness Through Physical Activity.* John Wiley & Sons.
70. Powers, S. K. and Howley, E. T. (2009). *Exercise Physiology: Theory and Application to Fitness and Performance, 7th Edition.* McGraw-Hill.
71. Reaburn, P., and Jenkins, D. (1996). *Training for Speed and Endurance*. Allyn and Unwin.
72. Riechman, S. E., Balasekaran, G., Roth, S. M., and Ferrell, R. E., (2004). Association of interleukin-15 protein and interleukin-15 receptor genetic variation with resistance exercise training responses. *Journal of Applied Physiology, 97*(6): 2214–2219.
73. Riechman, S. E., Zoeller, R. F., Balasekaran, G., Goss, F. L., and Robertson, R. J. (2002). Prediction of 2000 M Rowing Performance in Females Using Indices of Anaerobic and Aerobic Power. *Journal of Sports Science, 20*(9), 681–687.
74. Robertson, R. J. (2004). *Perceived exertion for practitioners: Rating effort with the OMNI picture system*. Human Kinetics.
75. Robertson R. J., Goss, F. L., Boer, N., Gallagher, J. D., Thompkins, T., Bufalino, K., Balasekaran, G., MeCkes, C., Pintar J., and Williams, A.

(2001). Omni Scale Perceived Exertion at Ventilatory Breakpoint in Children: Response normalized. *Medicine and Science in Sport and Exercise, 33*(11), 1946–1952.

76. Robertson, R. J, Goss, F. L., Boer, N. F., Peoples, J. A., Foreman, A. J., Dabayebeh, I. M., Millich, N. B., Balasekaran, G., Riechman, S. E., Gallagher, J. D., and Thomkins, T. (2000). Validation of the Omni Perceived Exertion Scale for Children Using A Mixed Gender/Race Cohort. *Medicine and Science in Sport and Exercise, 32*(3), 452–458.
77. Rowell, L. B. (1986). Circulatory adjustments to dynamic exercise. In *Human Circulation Regulation During Physical Stress*, pp. 213–256. Oxford University Press.
78. Rowland, T. W. (1996). *Developmental Exercise Physiology*. Human Kinetics Publishers.
79. Rushall, B. S. and Pyke, F. S. (1990). *Training for Sports and Fitness*. MacMillan Co.
80. Scott, W., Stevens, J., & Binder–Macleod, S. A. (2001). Human skeletal muscle fiber type classifications. *Physical Therapy, 81*(11), 1810–1816.
81. Spriet, L. L. (1995). Anaerobic metabolism during high-intensity exercise. *Exercise Metabolism*, 1–39.
82. Telford, R. D., Egerton, W. J., Hahn, A. G., and Pang, P. M. (1988). Skinfold measurement and weight control in elite athletes. *Excel, 5*(2), 21–25.
83. The Physical Fitness Specialist Certification Manual, The Cooper Institute for Aerobics Research, Dallas TX.
84. Thor, D. and Balasekaran, G. (2012). Comparison of cross-modal omni scale of perceived exertion at ventilatory breakpoint and self-regulated exercises in male adolescents in Singapore (Masters dissertation, National Institute of Education, Nanyang Technological University, Singapore).
85. Van Praggh, E. (1998). *Pediatric Anaerobic Performance*. Human Kinetics Publishers.
86. Wilmore, J. H. and Costill, D. L. (1988). *Training for Sport and Activity.* William C Brown Pub.

87. Wilmore, J. H. and Costill, D. L. (1999). *Physiology of Sport and Exercise*, 2nd edition. Human Kinetics, p. 310–341.
88. Wilmore, J. H. and Costill, D. L. (2005) *Physiology of Sport and Exercise*, 3rd edition. Human Kinetics
89. Wilmore, J. H., Costill, D. L., and Kenney, W. L. (1994). *Physiology of Sport and Exercise (Vol. 524)*. Human Kinetics.
90. Woodman, L. and Pyke, F. (1991). Periodisation of Australian football training. *Sports Coach, 14*(2), 32–39.

Index

CPSIA information can be obtained
at www.ICGtesting.com
Printed in the USA
FSHW021832040821
83631FS

9 789811 234156